Contents

Acknowledgements

The data on pesticides included in Part 4 are basically summarised from Martin and Worthing (1977) *Pesticide Manual* (5th edition), with supplementation from the 1977 edition of the M.A.F.F. *Approved Products for Farmers and Growers* for the cautionary comments, and in some cases the original data sheets from the chemical company concerned were used; these sources are gratefully acknowledged.

The publishers would also like to thank the following for providing photographs for the text: Ardea Photographics for figs. 1.6, 1.9, 1.13b; Biofotos for figs. 1.7a (J. Hodges), 2.9 (Heather Angel); J. Bridge for figs. 2.5a and b; Bruce Coleman Ltd. for figs. 1.7a, 2.2a; Centre for Overseas Pest Research for figs. 1.5a (N.D. Jago), 1.7d (R.E. Roome), 3.2 (P. Ellis), 4.3a (M.J. Hodson), 8.2 (D.J. McKinley). 9.1, 9.5 (E.S. Brown), 15.1a, 15.6a (P. Turner); T. Denham for figs. 14.4, 15.3, 15.4, 15.5; I.A.S. Gibson for fig 8.1; D.S. Hill for figs. 1.10, 6.5, 8.3, 8.5, 9.4; R.W. Hodgkiss for figs. 2.3a, 2.3b; ICI Plant Protection Division for figs. 1.3, 1.4, 1.5c, 1.5d, 1.7c, 1.8b, 1.12, 1.14, 2.2a, 2.2b, 3.3, 4.3b, 6.4, 6.7, 9.2, 9.3, 14.1, 14.3, 15.2; P. Jones for figs. 2.4a, 2.4b, 2.4c, 2.4d, 2.4e; G.A. Matthews for figs. 1.8a, 1.13a, 2.1, 4.1, 4.2, 15.1b, 15.6b, 15.7; R.I.S. for fig. 1.5b; Dr. F.M.L. Sheffield for fig. 6.3; Shell Photographic Library for fig. 1.5e; R.H. Stover for fig. 3.4; Tropical Products Institute for fig. 12.1; J.M. Waller for figs. 2.10a, 2.10b, 2.10c, 2.10d, 2.10e, 2.10f. 2.10g, 2.10h, 6.6, 6.8, 7.1, 8.4a, 8.4b, 12.2, 14.1, 14.2.

The publishers regret that they are unable to trace the copyright holders of the following photograph and apologise for any infringement of copyright caused: fig. 1.1.

The cover photograph was kindly supplied by ICI Plant Protection Division.

The authors acknowledge the help they received from many colleagues in the preparation of the text.

General Editor

Dr W.J.A. Payne
Consultant in tropical livestock production

Pests and Diseases of Tropical Crops

Volume 1: Principles and Methods of Control

D.S. Hill Ph.D., M. Sc., F.L.S., M.I. Biol.
Senior Lecturer in Entomology and Ecology
Department of Zoology, University of Hong Kong

J.M. Waller Ph.D., Dip. Agric. Sci., DIC, M.I. Biol.
ODA Plant Pathology Liaison Officer
Commonwealth Mycological Institute
Kew, Surrey, UK

Longman London and New York

Longman Group UK Limited,
Longman House, Burnt Mill, Harlow,
Essex CM20 2JE, England
and Associated Companies throughout the world

First published 1982
Fourth impression 1990

ISBN 0 582 60614 4

Produced by Longman Singapore Publishers Pte Ltd
Printed in Singapore

Library of Congress Cataloging in Publication Data

Hill, Dennis S., 1934-
Pests and diseases of tropical crops.

(Intermediate Tropical Agriculture Series)
Bibliography: v. 1, p:
Includes index.
Contents: v. 1. Principles and methods of control.
1. Tropical crops — Diseases and pests. 2. Pest contro
Tropics. I. Waller, J.M. II. Title. III. Series.
SB608.T8H54 632'.0913 81-18561
ISBN 0-582-60614-4 (v. 1) AACR2

Other titles in the Intermediate Tropical Agriculture Series

Already published:

H.T.B. Hall, *Diseases and Parasites of Livestock in the Tropics*
Describes the causes, symptoms, treatment and control of the main diseases of livestock in the Tropics.

L.R. Humphreys, *Tropical Pastures and Fodder Crops*
Describes natural tropical grasslands, discusses the possibilities for pasture improvement and suggests grass and legume species to accomplish this. Also includes sections on the establishment of pastures, soil fertility and pasture management.

J.C. Abbott and J.P. Makeham, *Agricultural Economics and Marketing in the Tropics*
Describes the inter-relations of agriculture, farm management, marketing and their economics, as they occur in the Tropics.

C.N. Williams and W.Y. Chew, *Tree and Field Crops of the Wetter Regions of the Tropics*
Details are supplied of the botany, climatic and soil requirements, cultivation and management, harvesting and, where appropriate, processing of a large number of crops.

J.A. Eusebio, *Pig Production in the Tropics*
Covers all aspects of pig raising in tropical areas, including nutrition, housing, breeding and marketing, with relevant biological details.

M.E. Adams, *Agricultural Extension in Developing Countries*
Explains the background and practicalities of extension work in the developing world.

H.F. Heady and E.B. Heady, *Range and Wildlife Management in the Tropics*
Covers all aspects of rangeland from planting and maintenance to cultural considerations.

D.V. Coy, *Accounting and Finance for Managers in Tropical Agriculture*
A useful guide to modern accounting practice for both students of agriculture and farm managers.

C. Devendra and G.B. Mcleroy, *Goat and Sheep Production in the Tropics*
Provides comprehensive coverage of how to rear and maintain healthy, productive goats and sheep in the Tropics. Includes sections on breeds, nutrition, reproduction, health and breed improvement.

Titles in preparation:

D.S. Hill and J.M. Waller, *Pests and Diseases of Tropical Crops: Volume 2, Field Handbook*

D. Gibbon and A. Pain, *Crops of the Drier Regions of the Tropics*

E. Heath and S. Olusanya, *Anatomy and Physiology of Tropical Livestock*

J. Ogborn, *Tropical Weeds*

C.N. Williams, *Growing Tropical Vegetables*

Glossary

acaricide Material toxic to mites (Acarina).
activator Chemical added to a pesticide to increase its toxicity.
active ingredient (a.i.) Toxic component of a formulated pesticide.
adherence The ability of a material to stick to a particular surface.
adhesive Material added to increase pesticide retention; different commercial preparations of methyl cellulose are available for this purpose.
adjuvant A spray additive to improve either physical or chemical properties (see also adhesive, emulsifier, spreader, sticker, supplement and wetter).
aedeagus The male intromittent organ, or penis, of insects.
aerosol A dispersion of spray droplets of diameter < 50 microns; usually dispersed from a cannister.
aestivation Dormancy during a hot or dry season.
aetiology The science and study of the factors causing disease.
agamic Parthenogenetic reproduction; non-sexual.
agitator A mechanical device in the spray tank to ensure uniform distribution of toxicant and to prevent sedimentation.
agroecology The study of ecology in relation to agricultural systems.
alternate host A plant which acts as sole host of a certain stage in the life cycle of a pest or pathogen.
alternative host A plant which acts as one of several hosts to a pest or pathogen.
anionic A negatively-charged molecule characteristic of certain wetting agents.
antagonistic (of micro-organisms) Reducing the growth or reproduction of pathogens when associated with them.
anthracnose A disease symptom characterised by dark sunken lesions.
antibiosis The resistance of a plant to insect attack by having for example, a thick cuticle, hairy leaves, toxic sap, etc.
antibiotic A substance produced by a micro-organism which kills or inhibits the growth of others.
anti-feedant A chemical possessing the property of inhibiting the feeding of certain insect pests.
anti-frothing agent Material added to prevent frothing of the liquid in a spray tank.
anti-sporulant A chemical which prevents a fungus from producing spores.
appressorium The swollen end of a fungal hypha which attaches it to a plant surface before penetration.
approved product (of pesticide) Proprietary brand of pesticide officially approved by the Ministry of Agriculture, UK.
arista A large bristle, located on the dorsal edge of the apical antennal segment in the Diptera.
ascomycete One of the major classes of fungi in which the sexual phase produces a number of spores (usually 8) in a sac called the ascus.
asexual Reproduction lacking the process of sexual fusion.
asymptote The point in the growth of a population at which numerical stability is reached.
atomiser Device for breaking up a liquid stream into very fine droplets by a stream of air.
atrophied Part of an organism which has not developed properly or has withered; reduced in size, rudimentary, vestigial.
attractant Material with an odour that attracts certain insects; lure; several proprietary lures are manufactured.
autocide A lethal substance produced within an organism which kills it; usually produced as a result of an induced genetic change; self-destructive method of pest control.
auxins Substances, usually produced within plants, which control plant growth; a plant hormone.
avoidance Disease control measures which rely on crops being grown in localities or seasons where or when a pest or disease is not active.

bacteria Single-celled microscopic plants which live as parasites or saprophytes.

bait Foodstuff used for attracting pests; usually mixed with a poison to form a poison bait.

band application Treatment of a band of soil in row-crops, usually covering plant rows, with either sprays or granules.

basidiomycete One of the major classes of fungi in which the sexual phase produces spores on stalks from a special cell called the basidium.

biocide Substance capable of killing a wide range of unrelated organisms.

biological control Control by the use of biological agents such as predators or parasites.

bionomics The study of organisms in relation to their environment, i.e. ecology.

biotic Biological; of living organisms.

bivoltine Having two broods or generations in a year.

black leg A disease symptom involving necrosis of the basal part of stems.

blast A plant disease symptom characterised by extensive death of aerial parts of the plant.

blight A plant disease symptom involving the death of leaves, flowers or stems.

boom (spray) Horizontal or vertical light frame carrying several spray nozzles.

brachypterous Having short wings, that do not cover the abdomen.

breaking (of emulsions) The separation of the phases from an emulsion.

budworm Common name in the USA for various tortricid larvae.

canker A disease symptom involving overgrowth of tissue at the edge of a lesion.

carrier Material serving as diluent and vehicle for the active ingredients of pesticides; usually in dusts.

caterpillar Cruciform larva; larva of a moth, butterfly, or sawfly, with many legs.

cationic surfactant Material in which surface activity is determined by the positively-charged parts of the molecule.

chaetotaxy The arrangement and nomenclature of the bristles on the insect's exoskeleton in both adults and larvae.

chemosterilant Chemical used to render an insect sterile without killing it.

chlorosis A disease symptom involving yellowing of leaves and other parts of a plant.

chrysalis The pupa of a butterfly.

coarctate pupa A pupa (Diptera) enclosed inside a hardened shell formed by the previous larval skin.

cocoon A protective case, often silken, inside which the larva turns into the pupa.

colloidal formulation Solution in which the particle size is less than 6 microns in diameter, and in which the particles stay dispersed.

compatibility The ability to mix different pesticides without physical or chemical interactions which would lead to reduction in biological efficiency or increase in phytotoxicity; or the ability of two organisms to live together in a balanced relationship, e.g. host and obligate parasite.

compressed (insects) Flattened from side to side.

concentrated solution (c.s.) Commercial pesticide preparation before dilution for use.

concentrate spraying Direct application of the pesticide concentrate without dilution.

concentration Proportion of active ingredient in a pesticide preparation, before or after dilution.

conidia Fungal spores produced asexually from specialised hyphae (conidiophores).

conidiophore A fungal hypha which produces conidia.

contact poison Material killing pests by contact action, presumably by absorption through the cuticle or skin.

continuous phase The liquid phase in a particulate suspension or the phase in which particles or droplets are dispersed in an emulsion.

control (noun) Untreated subjects used for comparison with those given a particular crop protection treatment.

control (verb) To reduce damage or pest density to a level below the economic threshold.
legislative The use of legislation to control the importation and to prevent any spread of a pest or disease within a country.

physical The use of mechanical (hand picking, etc.) and physical methods (heat, cold, radiation, etc.) of controlling pests or diseases.
cultural Regular farm operations designed to destroy pests or diseases.
chemical The use of chemical pesticides as smokes, gas, dusts and sprays to poison pests.
biological The use of natural predators, parasites and disease organisms to reduce pest populations.
integrated The very carefully reasoned use of several different methods of pest control in conjunction with one another to control pests with a minimum disturbance to the natural situation.

corm The swollen basal portion of a stem which acts as a perennating organ.

cosmopolitan A species occurring very widely throughout the major regions of the world.

costa A longitudinal wing vein, usually forming the anterior margin (leading edge) of the wing.

cover (of pesticides) Proportion of the surface area of the target plant on which the pesticide has been deposited.

climatograph A polygonal diagram resulting from plotting temperature means against relative humidity means.

crawler The active first instar of a scale insect.

cremaster A hooked, or spine-like, process at the posterior end of the pupa, often used for attachment (Lepidoptera).

crepuscular Animals that are active in the twilight, pre-dawn and at dusk in the evenings.

crochets Hooked spines at the tips of the prolegs of lepidopterous larvae.

crop rotation The successive growing of different crops on the same area of land.

cruciform larva Caterpillar; a larvae with a cylindrical body, well-developed head, and with both thoracic and abdominal prolegs.

cultivar A cultivated variety of a plant species.

curative The ability to cure (i.e. make healthy) a diseased plant.

damping off When seedlings collapse and die.

deflocculating agent Material added to a spray suspension to delay sedimentation.

defoliant Spray which induces premature leaf-fall.

deposit
spray Amount and pattern of spray or dust deposited per unit area of plant surface.
dried Amount and pattern of active ingredient deposited per unit area of plant surface.

deposition velocity Velocity at which the spray impinges on the target plants.

depressed (of insects) Flattened dorso-ventrally.

desiccant Chemical which kills vegetation by inducing extreme water loss.

desiccation Extreme drying usually causing death.

diapause A spontaneous state of dormancy or rest which some insects undergo in their life cycle.

diluent Component of spray or dust that reduces the concentration of the active ingredient but does not directly affect toxicity.

disease progress curve The result of plotting on a graph the increasing amount of disease in a crop against time.

disinfect To free from infection by destruction of a pest established in or on plants or plant parts; the removal of surface-contaminating micro-organisms especially pathogens.

disinfectant A substance which removes contaminant organisms from surfaces.

disinfest To kill or inactivate pests present upon the surface of plants or plant parts, or in the immediate vicinity (e.g. in soil).

dispersal Movement of individuals out of a population (emigration) or into a population (immigration); the movement of plant disease propagules.

dispersant A substance which facilitates the production of suspensions or emulsions in chemical sprays.

disperse phase The particulate phase of a suspension or the phase of an emulsion which is dispersed throughout the continuous phase.

diurnal Active during the daytime; a daily cycle.

dormant Alive but not growing; buds with an unbroken cover of scales; quiescent, inactive, a resting stage; ungerminated seeds or spores.

dose; dosage Quantity of pesticide applied per individual, or per unit area, or per unit volume, or per unit weight.

drift Spray or dust carried by natural air currents beyond the target area.

drop spectrum Distribution, by number or volume of drops, of spray into different categories of droplet sizes.

duster Equipment for applying pesticide dusts to a crop.

ecdysis The moulting (shedding of the skin) of larval arthropods between one stage of development and another; the final moult leading to the formation of the puparium or chrysalis.

ecoclimate Climate within the plant community.

ecology The study of all the living organisms in an area, their physical environment and interactions.

economic damage The injury done to a crop which will justify the cost of artificial control measures.

economic-injury level The lowest pest population density that will cause economic damage.

economic pest A pest causing a crop loss of 5–10 per cent according to definition.

economic threshold The pest-population level at which control measures should be started to prevent the pest population from reaching the economic-injury level.

ecosystem The interacting system of the living organisms in an area and their physical environment.

efficiency (of a pest control measure) The more or less fixed reduction of a population regardless of the number of pests involved.

effectiveness (of a pest control measure) This is shown by the number of pests remaining after control treatment.

elateriform larva A larva resembling a wireworm with a slender body, heavily sclerotinised, with short thoracic legs and only a few body bristles.

elytron or elytrum The thickened fore-wing of the Coleoptera.

emergence (of insects) The adult insect leaving the last nymphal skin, or pupal case. **(of seeds)** Germination of a seed and the appearance of the shoot.

emigration The movement of individuals out of a population.

emulsifiable concentrate (e.c.) Liquid formulation that when added to water will spontaneously disperse as fine droplets to form an emulsion.

emulsifier Spray additive which permits formation of a stable suspension of oil droplets in aqueous solution O/W, or of aqueous solution in oil, W/O.

emulsion, O/W A stable dispersion of oil droplets in aqueous solution.

emulsion, W/O Suspension of aqueous solution in oil.

encapsulation The encapsulation of a pesticide in a non-volatile envelope of gelatin, usually of minute size, for delayed release.

endemic (of pest or disease) Continuously present in an area.

entomophagous (of an animal or plant) An animal or plant which feeds upon insects.

environment Natural elements or objects surrounding an organism, e.g. air, soil, etc.

enzyme A protein produced by living cells which catalyses biochemical reactions such as digestion of food, etc.

epicuticle The outer layer of an insect cuticle.

epidemic An increase in disease incidence.

epidemiology The study of disease development and spread throughout a host population (such as a susceptible crop).

epidermis Outermost layer of cells covering leaves and young stems, etc. or insects.

eradicant A substance which eradicates a pathogen from the tissues of a host.

eradicate Completely remove or eliminate.

erineum A growth of hairs in dense patches on plant leaves resulting from the attack of certain gall-mites, Eriophyidae (Acarina).

exarate pupa A pupa in which the appendages are free and not fixed to the insect body.

exclusion Control of pests or diseases achieved by excluding them from an area or country, often as a result of phytosanitary legislation.

exotic Foreign; not indigenous or native.

extrinsic Factors or influences originating outside an organism.

exuvium The cast skin of arthropods after moulting.

facultative Optional, able to choose; not obligatory.
fallowing The practice of leaving agricultural land uncropped.
filler Inert component of pesticide dust or granule formulation.
flood fallowing The practice of flooding uncropped agricultural land with water usually for a period of several weeks or months.
flowability Property of flowing possessed by dusts, colloids, liquids and some pastes.
fluorescent tracer Fluorescent material added to a spray to aid the assessment of spray deposits on plants.
focus Specific area or site of initial disease development.
formulation (of pesticides) Statement of nature and amount of all constituents of a pesticide concentrate; or method of preparation of a pesticide concentrate.
fossorial Modified for digging; in the habit of digging or burrowing.
frass Wood fragments made by a wood-boring insect, usually mixed with the faeces.
fumigant Pesticide exhibiting toxicity in the vapour phase.
fumigation The application of gases or vapour to infiltrate soil, stored seeds or agricultural produce to kill pests and/or pathogens.
fungicide Substance capable of killing fungi.
fungistasis Prevention of the growth of a fungus without killing it.
fungistatic Preventing the growth of a fungus, but not killing it.
fungitoxic Poisonous (often lethal) to a fungus.
furrow application Placement of pesticides with seed in the furrow at the time of sowing.
gall An abnormal growth of plant tissues, caused by the stimulus of a pest or disease.
gene An heritable unit determining a character; a definite portion of the DNA in each cell.
generation The period from any given stage in the life cycle (usually adult) to the same stage in the offspring.
genetic Determined by genes; inherited.
germination Growth arising after a dormant state, e.g. of a seed or spore.
granulation The process of preparing substances (pesticides) as granules.
granule Coarse particle of inert material (pumice, Fuller's earth, rice husks) impregnated or mixed with a pesticide; used mainly for soil application, but sometimes for foliar application (pumice formulation).
granule applicator Machine designed to apply measured quantities of granules.
grease band Adhesive material (e.g. resin in castor oil, or 'Sticktite') applied as a band around a tree to trap or repel ascending wingless female moths, or ants.
grub (white) A scarabaeiform larva; thick-bodied, with a well-developed head and thoracic legs, without abdominal prolegs, usually sluggish in behaviour.
gustatory Applied to the sense of taste.
harrow Cultivation practice consisting of raking soil to produce a fine tilth.
haustoria Swellings or branches of fungal hyphae which penetrate host cells.
hemelytron The partly thickened fore-wing of Heteroptera.
hemimetabolous Insects having a simple metamorphosis, like that in the Orthoptera, Hemiptera, and Odonata.
herbivorous Feeding on plants; phytophagous.
heritable Capable of being passed from parents to offspring in successive generations.
hibernation Dormancy during the winter.
high-volume spraying see Spray.
hollow-cone Spray jet with a core of air, breaking to give drops in an annular pattern.
holometabolous Insects having a complete metamorphosis, as in the Diptera, Lepidoptera, Hymenoptera, Coleoptera.
homopterous Having the wings alike, of insects.
honeydew Liquid with high sugar content discharged from the anus of some Homoptera.
horizontal resistance Plant resistance operating generally against all races of a pathogen.
hornworm Larvae of Sphingidae with a dorsal spine or horn on the last abdonimal segment.
host The organism in or on which a parasite lives; the plant on which an insect feeds.

humectant Material added to a spray to delay evaporation of the water carrier.
hyaline Transparent.
hybridise To cross-fertilise two genetically distinct organisms.
hydrolyse The addition of a hydroxyl radical to a molecule.
hydrophillic Miscible with or soluble in water.
hydrophobic Immiscible with or insoluble in water, water-repellant.
hypermetamorphosis A type of complete metamorphosis in which the different larval instars represent two or more different types of larvae.
hyperparasite A parasite whose host is another parasite.
hyphae Individual filaments of a fungal mycelium.
hythergraph A polygonal diagram resulting from plotting temperature means against rainfall.
hypersensitivity Reaction of host cells to a parasite in which both are killed.
imago The adult, or reproductive stage of an insect.
immigration The movement of individuals into a population.
immune Exempt from infection by a pathogen; absolute resistance.
incompatible Not compatible; incapable of living together (as with host and parasite); incapable of forming a stable mixture with another chemical (as with pesticides).
indicator Marker.
indicator plant A plant which shows a characteristic reaction when infected with a virus.
indigenous Native, not foreign or exotic.
inert A material having no biological or chemical action.
infect To enter and establish a pathogenic relationship with a plant (host); to enter and persist in a carrier organism.
infection The process of, or state arising from, being infected (with a pathogen or parasite).
infest To occupy and cause injury to either plants, soils or stored products.
injector A device for positioning a pesticide below the soil surface or into the transport system of a tree.
inoculum That part or quantity of a pathogenic organism which can infect a host.
inoculum potential The quantity of a pathogen required successfully to infect a host, i.e. to overcome host resistance.
insecticide A toxin effective against insects.
instar The form of an insect between successive moults; the first instar being the stage between hatching and the first moult.
integrated control see Control.
intrinsic Factors or influences produced within an organism (e.g. as a result of its inheritance).
jet (of sprays) Liquid emitted from a nozzle orifice (in the USA, a nozzle).
knot A disease symptom involving gross malformation of plant tissue.
lacquer Pesticide incorporated into a lacquer or varnish to achieve slow release over a lengthy period of time.
latent Present but inactive; not apparent.
larva The immature stages of an insect having a complete metamorphosis, between the egg and pupa; the six-legged first instar of the Acarina.
larvicide Toxicant (poison) effective against insect larvae.
LC 50 Lethal concentration of toxicant required to kill 50 per cent of a large group of individuals of one species.
LD 50 Lethal dose of toxicant required to kill 50 per cent of a large group of individuals of one species.
leaf lamina The flat blade of a leaf.
leaf miner An insect which lives in and feeds upon the cells between the upper and lower epidermis of a leaf.
legislation The passing of laws to control activities.
lenticels Small pores in the bark of twigs which allow gaseous exchange.
lenticular Shaped like a double convex lens.
lesion Disruption (e.g. necrosis) of host tissue caused by a pathogen, or the toxic saliva of certain Heteroptera.

life table The separation of a pest population into its different age components.
lipophilic Soluble or easily miscible with oil.
localised Restricted to limited areas.
looper A caterpillar of the family Geometridae, with only one pair of abdominal prolegs (in addition to the terminal claspers), and which moves by looping its body.
low-volume spraying see Spray.
macropterous Large, or long-winged insect.
maggot A vermiform larva, legless, without a distinct head capsule (Diptera).
maturation Becoming mature, past the juvenile stage.
meristem Parts of plants which grow actively by cell division.
meristem tip culture Growing whole plants from small pieces of meristem tips of the parent plant.
microbial control Control of a pest by the use of micro-organisms which are usually antagonistic or pathogenic.
microclimate The conditions pertaining within the foliage canopy of plants; the climate within a microhabitat.
micro-encapsulation The incorporation of minute quantities of pesticide into extremely small capsules.
mildew Visible surface fungal growth on plants.
downy Diseases usually characterised by the production of a downy growth on the shoot surface; members of the fungal order Peronosporales.
powdery Diseases characterised by the production of a white powdery growth on the shoot surface; members of the fungal order Erysiphales.
miscible liquid (m.l.) A formulation of pesticide in which the technical product is dissolved in an organic solvent which is then on dilution dissolved in the water carrier.
mist blower Sprayer producing a fine air-carried spray.
miticide see Acaricide.
molluscicide Toxicant effective against slugs and snails.
monoculture The extensive cultivation of a single species of plant.
monophagous An insect restricted to a single species of host plant.
mortality Death-rate.
mosaic Mottled pattern on leaves affected with certain virus diseases.
motile Moving; capable of movement.
mould Microfungus, the delicate mycelial growth produced by many fungi.
multiline A crop, or 'composite cultivar', consisting of several genetically distinct groups or 'lines' of a cultivar.
mutation A sudden genetic change, producing a different heritable character.
mycelium The filamentous mass of hyphae constituting the vegetative body of a fungus.
mycoplasma Minute intercellular pathogen.
natality Birth-rate.
native Indigenous, originating locally, not exotic.
necrosis Death of part of a plant.
nematicide Toxicant effective against nematodes (eelworms).
nocturnal Active at night.
non-ionic surfactant A surfactant that does not ionise in solution and is therefore compatible with both anionic and cationic surfactants.
non-persistent
chemical Chemical that remains active for only a short period after application.
virus A virus that remains infective for only a short period after acquisition by a vector.
nozzle Attachment for spraying liquids.
air blast Nozzle using high-velocity air to break up the spray liquid supplied at low pressure.
anvil Nozzle in which the spray liquid jet strikes smooth solid surface at a high angle of incidence.
cone or swirl Nozzle in which the liquid emerges from the orifice with tangential velocity imparted by passage through one or more tangential or helical channels in the swirl chamber.
hollow cone Nozzle in which spray jet has a core of air, breaking to give drops in annular pattern.

fan nozzle The aperture is an elongate horizontal slit, producing a fan-shaped spray pattern.

deflector Nozzle in which a fan-shaped sheet of spray is formed by directing the liquid over a sharply inwardly-curving surface.

nymph The immature stage of an insect that does not have a distinct pupal stage.

obligate Compulsory, without choice; not facultative.

obtect pupa A pupa in which the appendages are more or less fixed to the body surface as in Lepidoptera.

oligophagous An animal feeding upon only a few, closely-related host plants, or it may be an animal parasite.

onisciform larva A flattened platyform larvae like a woodlouse in appearance.

orifice (nozzle) velocity Velocity at which the spray leaves the nozzle orifice.

ovicide Toxicant effective against insect or mite eggs.

oviparous Reproduction by laying eggs.

oviposition Laying of eggs.

paint gun Type of small, air-blast machine.

pan-tropical A species occurring widely throughout the Tropics and sub-tropics.

parasite An organism living in intimate association with a living organism (plant or animal) from which it derives material essential for its existence, while conferring no benefit in return.

parthenogenesis Reproduction without fertilisation; usually through eggs but sometimes through viviparity.

parts per million (ppm) A measurement of concentration, e.g. proportion of toxicant present, in relation to that of plant material on which it has been deposited; usually used in connection with the edible portion of a crop and its suitability for consumption.

pathogen A parasite causing disease symptoms in its host.

pathogenesis The production of disease symptoms in the host by a pathogen.

pathotype A sub-division of a species based on differences in pathogenic characteristics.

pellet Seed coated with inert material, often incorporating pesticides, to ensure uniform size and shape for precision drilling.

penetrant Oil added to a spray to enable it to penetrate the waxy insect cuticle effectively.

persistent (of pesticide) The term applied to chemicals that remain active for a long period of time after application.

(of virus) A plant virus which remains infective with its vector for a long time, non-persistent viruses quickly lose their infectivity.

pest An organism, usually an animal, causing damage to man's crops, animals or possessions; sometimes used in the wider sense to include diseases as well.

pest complex The group of pests attacking a crop at any one time.

pest density The population level at which a pest species causes economic damage.

pest management The careful manipulation of a pest situation, after extensive consideration of all aspects of the life system as well as ecological and economic factors.

pest spectrum The complete range of pests attacking a particular crop.

pesticide A toxic (poisonous) chemical used to kill pest organisms; a term of wide application which includes all the more specific applications — insecticide, acaricide, bactericide, fungicide, herbicide, molluscicide, nematicide, rodenticide, etc.

pheromone Ectohormone; a substance secreted by an insect to the exterior causing a specific reaction in the receiving insects.

phloem Conducting tissue in plants which conveys organic substances in solution.

phytoalexin A substance, toxic to an infecting pathogen, produced during the hypersensitive reaction of host cells.

phycomycetes A major class of fungi.

phytophagous Herbivorous; plant eating.

phytosanitation Measures requiring the removal or destruction of diseased or infested plant material likely to form a source of re-infection or re-infestation.

phytotoxic A chemical liable to damage or kill plants (especially crop plants), or plant parts.

planidium larva A type of first instar larvae in certain Diptera and Hymenoptera which undergo hypermetamorphosis.

poison bait An attractant foodstuff for insects, molluscs or rodents, mixed with toxicant.

polyphagous An animal feeding upon a range of hosts.

pre-access interval The interval of time between the last application of pesticide to an area and safe access to the area for domestic livestock and persons without protective clothing.

predator An animal which preys upon others.

predispose Favour; make more susceptible.

predisposition Making a plant more susceptible to a pest or disease — usually as a result of environmental or cultural defects.

preference The factor by which certain plants are more or less attractive to insects by virtue of their texture, colour, aroma or taste.

pre-harvest interval The interval of time between the last application of a pesticide and the safe harvesting of edible crops for immediate consumption.

pre-oviposition The period of time between the emergence of an adult female insect and the start of its egg laying.

prepupa A quiescent stage between the larval period and the pupa; found in some Diptera and Thysanoptera.

preventative A measure applied in anticipation of pest or disease attack.

proboscis A trunk-like process of the head of some insects (Lepidoptera) used for feeding.

progeny Offspring; second generation seedlings resulting from parent plant.

proleg A fleshy abdominal leg found in caterpillars (Lepidoptera, sawflies).

propagate To reproduce, increase numbers of individuals by sexual, asexual, or vegetative means, e.g. rooted cuttings.

propagule Part of an organism capable of propagating the species, e.g. seed, spore, etc.

proprietary name Manufacturer's name or brand name given to a particular formulated product.

protectant A substance, e.g. a fungicide, conferring protection (against a pathogen) on a plant.

protection The means taken to control a pest or disease on a given crop.

protective clothing Clothing to protect the spray operator from the toxic effects of crop protection chemicals, this may include rubber gloves, boots, apron, respirator, face mask, etc.

protonymph The second instar of mites.

pterostigma A thickened opaque or dark spot along the costal margin of the wing, near the tip (Odonata, Hymenoptera, etc.)

pupa The stage between larva and adult in insects with complete metamorphosis; a non-feeding and usually inactive stage.

puparium The case formed by the hardened, last larval skin in which the pupa of some Diptera is formed.

pupate The stage in the life cycle of an insect when the larva becomes the pupa.

quarantine All operations associated with the prevention of importation of unwanted organisms into a territory, or their exportation from it.

race (of a pest or pathogen) Subdivision of a species characterised by pathogenic properties, e.g. biotype, pathotype, etc.

rate of oviposition Rate of laying eggs.

redistribution (of pesticide) Movement of pesticide subsequent to initial application to other parts of the plant, usually by rain.

repellant A chemical which has the property of inducing avoidance by a particular pest.

residual poison Poison remaining in the pest's body for some time after application and still capable of harming the pest.

residue Amount of pesticide remaining in or on plant tissue (or in soil) after a given time, especially at harvest time.

resistance The natural or induced capacity of a plant to avoid or repel attack by pests or parasites; the ability of a pest to withstand the toxic effects of a pesticide or a group of pesticides, often by metabolic detoxification.

rhizomorph A root-like growth of fungal mycelium, e.g. in soil.

rhizosphere The zone of active microbial growth around the root.
rodenticide A toxicant effective against rodents.
rogueing The removal of unhealthy or unwanted plants from a crop.
rostrum The beak or stylet of Hemiptera.
rot A disease symptom in which plant tissues are destroyed.
run-off The process of spray shedding from a plant surface during and immediately after application, when droplets coalesce to form a continuous film and surplus liquid drops from the surface.
rust A type of disease characterised by the production of yellow to red pustules of spores on the surface of the host.
sanitation Removal of diseased material; reduction in inoculum; phytosanitation with reference to plant material.
saprophyte An organism living on dead organic matter.
scarabaeiform larva A grub-like larva, with a thickened cylindrical body, well-developed head and thoracic legs, without abdominal prolegs, and sluggish in behaviour.
scavenger An animal that feeds on dead plants and animals, on decaying matter, or on animal faeces.
sclerotia Tough resting bodies produced by some fungi.
sedimentation The deposition of particles from suspension.
seed dressing A coating, either dry or wet, of protectant pesticide applied to seeds before planting; dry seed dressings are often physically stuck to the testa of the seed by a sticker such as methyl cellulose.
semi-looper Caterpillar from the sub-family Plusiinae (Noctuidae) with two or three pairs of prolegs which locomotes in a somewhat looping manner.
shifting cultivation Growing crops on land reclaimed from bush for a few years and then moving on to a new area, the old area reverting again to bush.
siphunculi The paired protruding organs at the terminal end of the abdomen of Aphidoidea, also called cornicles.
slurry Paste-like liquid used as a seed coating.
smoke Aerial dispersal of minute soil particles or pesticides through the use of combustible mixtures.
smut A type of disease in which a part of the host is converted into a mass of black spores.
soil sterilant Toxicant added to, or injected into, soil for the purpose of killing pests and pathogens.
solid cone Jet with air cone reduced to give a cone of spray droplets.
solvent Carrier solution in which the pesticide is dissolved to form the concentrate.
spore A fungal propagule, initially dormant and often specialised for dispersal, survival, etc. in different ways.
sporulation The process of producing spores.
spray
- **air-carried** Spray propelled to target in a stream of air.
- **coarse** Dispersion of droplets of mass median diameter over 200 microns.
- **concentrate** Undiluted commercial pesticide preparation.
- **fine** Dispersion of droplets of mass median diameter from 50 - 150 microns.
- **floor** Applied to the litter on the ground surface.
- **high-volume** Over 1 000 litres per hectare on bushes and trees; over 700 litres per hectare on ground crops (or over 400 litres per hectare according to definition).
- **low-volume** Spray of 200 - 600 litres per hectare on bushes and trees; 50 - 200 litres per hectare on ground crops or 5 - 40 litres per hectare according to definition.
- **median-volume** 600 - 1 100 litres per hectare on bushes and trees; 200 - 700 litres per hectare on ground crops.
- **mist** Dispersion of droplets of 50 - 100 microns in diameter.
- **ultra-low-volume** Less than 50 litres per hectare on ground crops; less than 200 litres per hectare on trees and bushes; or less than 0·5 - 5 litres per hectare, according to definition.

spray angle Angle between the sides of a jet leaving the orifice.

sprayer Apparatus for applying pesticide spray; not to be confused with 'spray operator'.
spread (of pesticide) Uniformity and completeness with which a spray deposit covers a continuous surface, such as a leaf or a seed.
spreader Material added to a spray to lower the surface tension and to improve spread over a given area; wetter.
spur An articulated spine, often on a leg segment, usually the tibia; a serrulate tibial spur is characteristic of the Delphacidae (Homoptera).
stability The ability of a pesticide formulation to resist chemical degradation over a period of time, either in storage or after application.
sterilant A substance used to sterilise.
sterile Devoid of living organisms; infertile.
sterilisation To make sterile by killing any micro-organisms present; to render infertile.
sticker A material of high viscosity used to stick powdered seed dressings on to seeds; two commonly used stickers are paraffin (kerosene) and methyl cellulose; a solution of methyl cellulose can be added to a spray to increase retention on plant foliage.
stimulus A substance or force producing a reaction from a plant or animal.
stomach poison A toxicant (poison) which operates by absorption through the intestine after having been ingested by the insect, usually on plant material.
stomata Microscopic pores in leaves and young stems through which water evaporation (transpiration) and gaseous exchange occurs.
stylet The piercing beak or mouthparts of sap-sucking bugs (Hemiptera).
suberisation The production of corky tissue in plants, as in scabs.
susceptible Capable of being infected, easily becoming diseased; not resistant.
supplement (spray) Adjuvant; additive.
surfactant Spreader; wetter.
swath Width of target area sprayed at one pass.
symptom Visible signs of disease or pest attack.
syndrome A group of concomitant symptoms, often characteristic of a particular disease or pest attack.
synergism Increased pesticidal activity of a mixture of pesticides above that of the sum of the values of the individual components; or increased virulence of a mixture of pathogens and/or pests.
synergistically To act together to produce synergism.
systemic A pesticide absorbed through the plant surfaces (usually root) and translocated through the plant vascular system; a symptom involving the whole of the plant; a pathogen present in all or most of the plant tissues.
taint Unwanted flavour in fresh or processed food from a pesticide used on the growing crop.
target surface The surface intended to receive a spray or dust application.
technical product The usual form in which a pesticide is prepared and handled prior to formulation; usually at a high level of purity (95–98 per cent) but not completely pure.
tegmen The thickened and leathery fore-wing in the Orthoptera and Dictyoptera.
tenacity The property of a pesticide deposit or residue which enables it to resist removal by weathering.
tenacity index Ratio of the quantity of residue per unit area at the end of a given period of weathering to that present at the beginning.
therapeutants Chemicals which have a curative action on diseases.
therapeutic Curative, capable of eradicating or reducing the effect of a disease.
tissue The mass of cells which make up plant organs.
tolerance Ability to endure infestation (or infection) by a particular pest (or pathogen) without showing severe symptoms of distress or marked loss of yield.
tolerance, permitted Maximum amount of toxicant allowed in foodstuffs for human or animal consumption.
topography The structure of land, e.g. slope.
toxic Poisonous.
toxicant Poison; chemical exhibiting toxicity.
toxicity Ability to poison or to interfere adversely with the vital processes of the organism.

toxin Substance detrimental to an organism; a poison.
tracer A radioactive or fluorescent molecule used to locate a deposit.
translaminar A pesticide which passes through the leaf tissue from one surface of a leaf to the other (from lamina to lamina).
translocation The uptake of a pesticide into part of a plant body and its subsequent dispersal to other parts of the plant.
trap crop A crop, sometimes of wild plants, grown especially to attract pests or diseases and when infested (or infected) either sprayed or collected and destroyed; trap plants are usually grown between the rows of the crop plants or around the edges of fields.
triungulin larva The active first instar larva of Meloidae (Coleoptera) and Strepsiptera.
tuber A swollen underground stem acting as a perennating or storage organ e.g. potato.
tylose A swelling or bulging of cell walls blocking the xylem elements of woody plants, often as a response to infection.
ultra-low-volume spraying (ULV) see Spray.
univoltine Producing one brood or generation in a season.
uredospore Asexual spore produced by rust fungi.
vascular Pertaining to conducting tissues.
vector Organism able to transmit viruses or other pathogens either directly or indirectly; direct virus vectors include insects, mites and nematodes.
vermiform larva A legless (apodous), headless (acephalic), worm-like larva typical of some Diptera.
vertical resistance Plant resistance to disease only effective against some races of a pathogen; resistance capable of being breached by new races (pathotypes) and therefore usually temporary.
vestigial Poorly developed; degenerate; non-functional.
viable Alive, even if in a dormant state, and capable of germination.
viviparous Giving birth to living young (e.g. Aphidoidea).
virulence A measure of the ability of a pathogen to infect a host and cause disease.
virus Minute intercellular disease agent.
volunteer Crop plants growing accidentally from shed seeds; without intentional cultivation.
wettable powder (w.p.) A formulation of a pesticide in powder form; easily miscible with water in which it forms a suspension.
wetter A substance which decreases the surface tension of a liquid spray; additive.
wilt A disease symptom involving collapse of plant tissues.
wireworm The larva of Elateridae (Coleoptera); long, slender, well sclerotised, thoracic legs but not prolegs, and few setae.
witch's broom A symptom of disease characterised by the profuse production of lateral shoots.
xylem Water-conducting tissue in plants.

Introduction

Insects cause the greatest hazard to man's use of the world's vegetation. There are more than 1 million different species of insects and they are found in all parts of the world in every possible habitat. Many species produce huge numbers of individuals. Most insect pests are herbivorous and the insect attacks all parts of the plant structure. It is probable that 10 000 of these species are agricultural pests of some consequence. Some major crops such as coconut, coffee and cotton, may be damaged by as many as 500–700 different species of insect. In general, the number of major insect pests of any crop is between 5 and 20.

Crops are attacked also by plant diseases. There is a close relationship between these diseases and insect pests. Insects may themselves be transmitters of diseases or may so weaken the plants by their attack that diseases can flourish.

Plant diseases may be caused by fungi, bacteria or viruses. About 8 000 plant pathogens have been recorded many of which attack several crops. Nematodes can also be pathogenic. Immense losses may be caused by severe epidemics of diseases of major crops. Examples of these and the effects that have resulted are: potato blight which was responsible for the Irish famine, coffee rust which destroyed the coffee industry of Sri Lanka, and wheat rust which caused successive disruptions to American cereal production. Less attention has been paid to losses caused by diseases which have less dramatic effects. Some diseases maintain a low level of attack and do not reach epidemic proportions. These can, however, cause heavier losses over wider areas and for longer periods than more spectacular diseases.

Insect pests may reinforce the effects of plant disease and vice versa. About 15 per cent of the world's total crop is lost during cultivation as the result of pest attack and a further 20 per cent is lost during post-harvest storage. In most cases the loss in tropical countries is greater than the world average and it is probable that 20 per cent of the total crop is lost during cultivation and a further 30 per cent during storage. In developed countries, it is estimated that disease causes a 10 to 20 per cent overall loss in crops so that in the Tropics a loss of 15 to 30 per cent may be estimated. The combined effect of pest attack and plant disease may result in a loss of 35 to 50 per cent of the potential yield of crops in tropical areas. Pest attacks may cause a further 20 per cent loss during post-harvest storage of the crops.

These losses are often greatest in the countries that can least afford them. A reduction of these losses, with a resulting increase in agricultural productivity is therefore of paramount importance to developing countries in the Tropics. The application of proper control measures can have spectacular results. For example, it has been reported from Ghana that since the control of cocoa capsid bugs the yield of cocoa has trebled. Achievement of control can only be accomplished by cooperation between research and extension workers and the farmer who has finally to apply the control measures that may be required.

A great deal of research work on the control of insect pests and diseases has already been carried out. The knowledge gained must now be made available to the farmer and applied at the practical level.

It is usual for applied entomologists and plant pathologists to write separate texts on insect pests and plant disease control. It is, however, the overall health of plants and thus the combined effect of pests and diseases on a specific crop which concerns the farmer and extension workers. We hope that this text and the companion volume entitled *Pests and Diseases of Tropical Crops Volume 2 Field Handbook* will be particularly suited to the practical needs of agricultural students, extension workers and farmers in developing tropical countries.

There are of course many excellent specialised texts and monographs on the pests of specific crops, on pesticides, on certain groups of insects of economic importance, on the diseases of crops and on various aspects of plant pathology, that deal with these subjects in much greater detail than do the two texts in this series. What is intended, is that readers should gain a basic appreciation and knowledge of the need for crop protection to help them recognise pest and disease problems when they occur. It is hoped that this will stimulate an interest in the more specialised texts and draw the attention of appropriate specialists to the many difficult problems in the field of tropical crop protection.

Further Reading

Fletcher W.W. (1974). *The Pest War*. Blackwells: Oxford.

Part 1 Principles of pest and disease control

1 Pests

Definition of a pest

The definition of a pest can be extremely subjective, varying according to many factors, but in the widest sense any animal or plant which harms or causes damage to man, his animals, crops, or possessions, or even just causes him annoyance, qualifies for the term **pest**.

Self-sown or volunteer plants out of context can be regarded as pests; thus volunteer cereals in a field growing sweet potato have to be treated as weed pests. On an agricultural basis a pest is that which causes a loss in yield or quality of the crop resulting in a loss of profits by the farmer. When a loss in yield reaches certain proportions then the pest can be defined as an **economic pest**. It is accepted that **pest status** is reached when there is a 5 to 10 per cent loss in yield in a particular crop. **Economic damage** is the amount of injury done to a crop which will justify the cost of artificial control measures. The **economic injury level** is the lowest pest population density that will cause economic damage to a crop and this will vary from crop to crop, season to season, and area to area. For practical purposes there is an **economic threshold** which is defined as: the population density at which control measures should be started to prevent an increasing pest population from reaching the economic injury level. In many cases a particular crop is attacked by a **pest complex** of insects and diseases.

The control of a pest complex presents various complications since several different methods of control may be required, this makes evaluation of the requirements difficult.

Key pests

It is usually possible to single out one or two pests in the complex as most important. These can be defined as **key pests**. They are perennial and usually have high reproductive potential. Their importance is often due to the type of damage suffered by the crop. Key pests dominate control practices.

A single crop may have one or more key pests which may vary in different areas and from time to time. Variation occurs because their numbers are controlled by complex biotic and physical factors.

When individuals of a key pest occur on a crop, a proper investigation should be carried out before control measures are applied. Determination of the economic threshold avoids unnecessary application of pesticide.

Ecosystems

The earliest recorded attempts at pest control attempted to make the environment less favourable for the pests by cultural and biological means. The concern lay with the biology of the pests and their ecology. The **ecosystem** is the complex interacting system of all the living organisms, plants and animals, of an area and their physical environment, climate, soil, water, shelter, etc. **Ecology** is the study of ecosystems and can be defined as the total relationships of animals and plants, to each other, and to their environment.

Environment The environment comprises four main factors: weather; food; other animals and plants; shelter, a place in which to live.

Environmental factors may be regarded as biotic, i.e. organic, and physical or abiotic i.e. inorganic factors. Weather and shelter are physical factors whereas other animals and plants and sometimes shelter are biotic factors. Food is a biotic factor for animals which are holozoic (heterotrophic) in their feeding habits, but could be more suitably described as physical for plants, which are holophytic (autotrophic) in their nutrition.

The environmental factors include the following aspects.

1 *Weather*: includes consideration of temperature, humidity, water (rainfall, ground water, etc.), light and wind.
2 *Food*: (a) animals: organic remains (detritivores); plants (herbivores); other animals (carnivores and parasites);
(b) plants: organic remains (saprophytes); other plants and animals (parasites and insectivorous plants); sunlight, water, carbon dioxide, chlorophyll, etc. (autotrophs — typical plants).
3 *Other animals and plants*: competition — interspecific (between different species) and intraspecific (within the species); predation; parasitism; pathogens causing diseases.
4 *Shelter*: (a) for insects (and pathogens) it is frequently a plant, and often a specific location on the plant, e.g. leaf miner, stem borer, boll borer, pod borer, leaf roller, etc. (Fig. 1.1).

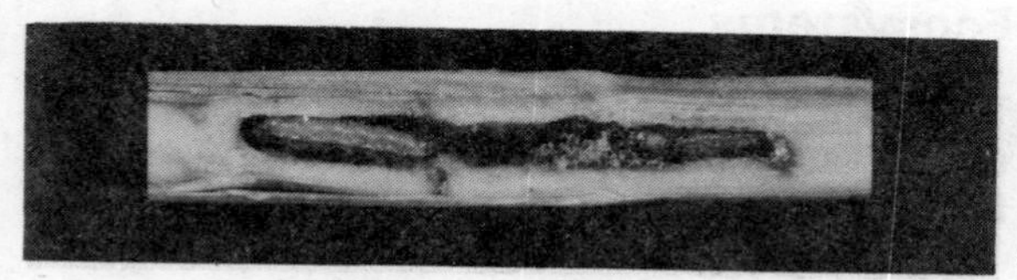

fig. 1.1 These caterpillars of the stem-borer *Busseola fusca* sp. shelter in the maize stem they have eaten

(b) for plants it is a physical location (including the soil) together with the other plants that constitute the community.

Pest management The disastrous effects of widespread and continual use of chemical pesticides, especially the early organochlorine compounds, led to general disillusionment with the method. The contemporary approach is one of **integrated control** or **pest management** (p. 65). This is a return to the traditional concern with biological and ecological understanding as a basis for the correct use of selected pesticides.

The interaction of a pest and its environment is complex. Thus the possibilities for ecologically orientated control are considerable. To give the best chance of successful control a detailed knowledge of the insect's life history and biology and especially its relationship with the host plant is essential.

In a pest ecosystem one or more aspects of the environment may be of overriding importance. The key to control will inevitably lie in the understanding of the complex of environmental factors and their relative importance. However, at present our knowledge of most pest ecosystems falls far short of this ideal and a great deal of basic ecological study is still required. Too frequently pest control still consists of hasty applications of chemical pesticides to which little consideration has been given. This often fails to control the pest and may well cause immense ecological damage to the area.

Agroecosystems

An agroecosystem is basically the ecosystem of an area as modified by the practice of agriculture, horticulture or animal rearing. Agriculture consists of methods of soil management and plant cultivation designed to achieve a maximum sustained yield of crop produce in the shortest possible time. To do this changes are made in the environment to improve growing conditions for the plants and to minimise damage by pests and diseases. These include irrigation, use of fertiliser, provision of shelter-belts, mulching and crop-spacing as well as the application of herbicides, fungicides, insecticides and other pesticides as required.

The most serious agricultural modifications are probably those concerned with monoculture which includes the simplification of the flora and lessening of competing species. The crop plants are more succulent than their wild relatives and are thus more attractive to the pest. In addition the agricultural microenvironment and microclimate are more uniform. For further discussion of these points see p. 21.

Pests in agroecosystems are also affected by changes in the usual predator/prey relationships; the introduction of exotic crop plants, weeds and pests; and new techniques such as minimal cultivation.

In addition agricultural practices change for economic and other reasons, for example, with many crops it is more economical for them to be grown in single large areas permitting easy access and manipulation by machinery. Thus the hedgerows and strips of wild vegetation between adjacent fields are destroyed in the production of larger more uniform fields. In the Tropics this development is not yet extensive but in most European countries the destruction of traditional hedgerows is both widespread and intensive. This practice is generally more advantageous than deleterious to farming, although local floral and faunal impoverishment inevitably results.

Pest populations in the Tropics

The technologically advanced countries have the resources to develop adequate procedures for control but it is now realised that these cannot be applied uncritically in tropical conditions.

In temperate regions the life cycle of most species is regular, usually following seasonal changes in temperature. Control measures based upon detailed knowledge of the species' life history, if applied with appropriate timing, can be most effective. For example, a population of caterpillar pests can be sprayed with insecticide when most individuals are in the young, susceptible, larval stage, thus achieving a high kill rate. The use of other insects for natural control by predation and parasitism is not important in temperate climates. There is rarely enough time in the short growing season for attacking populations to build up sufficient numbers to control the pest. However, predation by non-specific predators can be significant.

In many parts of the Tropics, particularly in tropical rain-forests, the climate tends to be warm with little fluctuation of temperature either daily or seasonally; rainfall, relative humidity, and other environmental factors also vary little. So most pests are constantly breeding. All developmental stages are present most of the time and insecticide sprays are less effective since the eggs and pupae generally survive contact with the poison. Under tropical conditions translocation and degradation of the insecticide are more rapid so that spray treatments are generally less effective.

In these climatic conditions phytophagous insects may be controlled by predators, pathogens, and parasites. Many tropical pest species are normally restrained by their natural enemies; any restriction of the attacking population can cause disastrous increases in numbers. The full extent of natural control has only been investigated in a few cases. Typically it is only appreciated after pesticide disasters, where the broad-spectrum, persistent, contact insecticides have depleted the natural predators and parasites more than the pest population. Natural enemies attack all stages of insects and in the few cases studied, mortality from egg to adult amounts to 90–99 per cent. Two well-documented examples of ill-advised spraying with insecticide are recorded from Malaysia. Serious pest outbreaks on oil palm and cocoa followed extensive application of persistent insecticides against more minor pests. Large areas of these crops were damaged. The pest outbreaks were due to the destruction of natural parasites and to other factors resulting from the ecological chaos in the plantations caused by the insecticides. The breakdown in natural synchronisation of the life cycles of the pests and their natural enemies probably played an important part.

Perennial crops maintain continuity over a period of years and are more subject to natural control of pests. With annual or short-lived crops conditions resemble those of the temperate crops where there has to be an initial infestation, controlled in this instance by availability of host plants rather than suitable climatic conditions. So the short-term tropical crops are more subject to pest outbreaks since periods of great suitability alternate with periods of total unsuitability for particular pests. This state is highly disruptive of biological balance and likely to lead to population upsurges of phytophagous insects.

Population level

A pest is only an economic pest at or above a certain population density. The control measures employed against it are designed usually only to lower the population below that density; the aim is rarely complete eradication of the pest. Fig. 1.2 is a schematic representation of the growth of a population; it shows 4 separate population levels, represented by the numbers 1 to 4. These population levels indicate hypothetical densities at which any particular pest species may be designated an

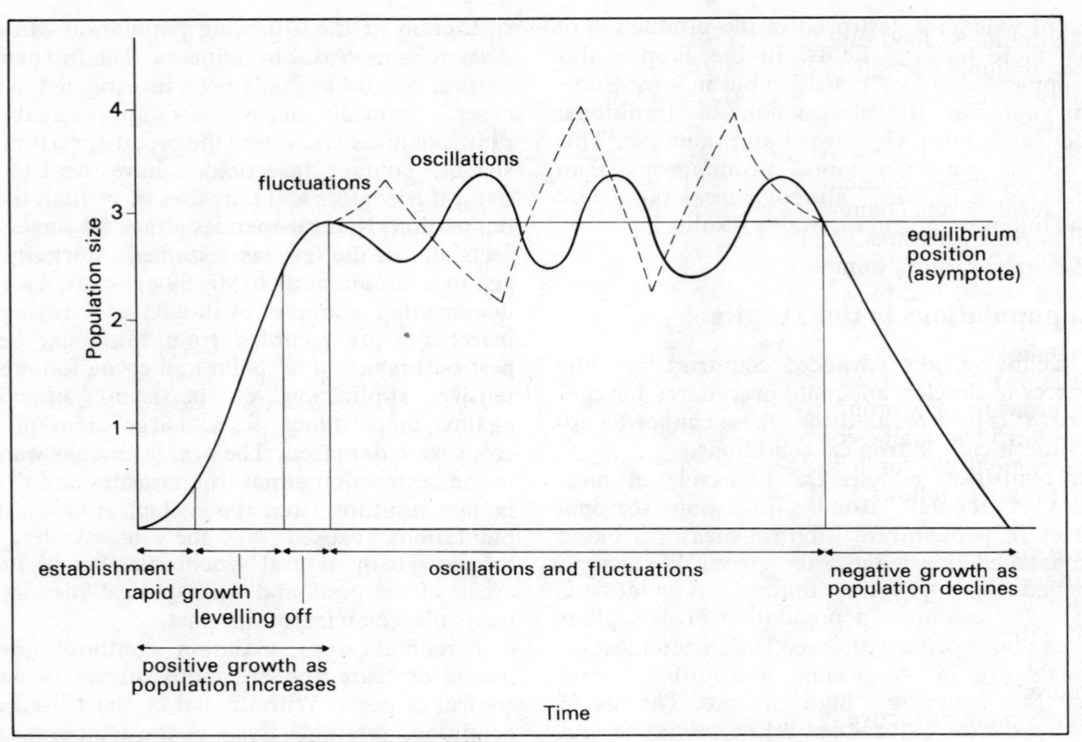

fig. 1.2 The growth of populations (*Source*: Allee, (1950))

economic pest. Population level 1 might well represent the economic pest level for a heteropteran bug, e.g. *Antestiopsis* spp. In this instance control measures are recommended when the population density reaches 2 bugs per bush, whereas observations on isolated unsprayed arabica coffee bushes in Uganda indicate that the average pest level may be in the region of 20–40 bugs per bush. At the other extreme, population level 4 could well apply to pests such as the desert locust which is only an economic pest in Africa at irregular intervals, at times of population eruption. Most of the more common pests would come into the categories which reach pest density at population levels about 2 and 3. The growth of a population can be expressed very simply in the following equation:

$$P_2 \rightleftharpoons P_1 + N - M \pm D$$

where P_2 = final population; P_1 = initial population; N = birth rate; M = death rate; D = dispersal.

Dispersal is regarded as either movement out of the population, **emigration**, or movement into the population from outside, **immigration**. The examination of a pest population and its separation into the separate age-group components, i.e. eggs, larvae, pupae, adults, enables a **life table** for that pest population to be compiled. A study of the life-table of a pest reveals the dynamic state of that population and its probable importance as a pest.

The object of most pest control measures is to lower P_2, which can be done by either lowering the birth rate of the pest, increasing the death rate, or inducing the pest to emigrate away from the area concerned.

Development of pest status

Increase in numbers Pest status is most commonly attained simply by an increase in numbers. The natural control of a population is upset by the practice of agriculture when the planted crop provides an unlimited food supply for a potential pest. The population may still be kept in check by its predators and parasites but usually the natural control factors do not act quickly enough. When this happens the farmer must use control methods to avoid crop losses. Seasonal increases in number usually result from changing climatic conditions and biological pressures. The climatic conditions include temperature, humidity and rainfall; the biological ones include competition, both intra- and interspecific, parasitism and predation.

Ecological change A harmless species may be converted into a pest by an ecological change such as the growing of a crop which is susceptible to that species. The major ecological changes which can be responsible for a species developing into pest status are as follows.

1 *Character of food supply* Plants grown for agriculture have usually been selected for their nutritive value and are typically large and succulent, with especially large fruits or seeds. Thus maize and sorghum are far more attractive food to stalk-borer caterpillars than are wild grasses; and cabbage is more attractive to caterpillars (Fig. 1.3) than wild crucifers.

2 *Monoculture* A particular species of crop may not always be grown on exactly the same site, but even in a rotation it may be in fields adjacent to the original site and open to attack by mobile pests. Monoculture is essentially similar to the climax conditions of some natural temperate vegetation, where large areas are dominated by very few plant species. The devastating attacks sometimes recorded on forest trees in North America and Europe by defoliating caterpillars is the natural equivalent of a field crop heavily infested with insect pests.

3 *Minimum cultivation techniques* Minimum cultivation is a recent agricultural technique in soil preparation. It consists essentially of a chemical destruction of old crop remains and weeds followed by planting of the new crop into the undisturbed soil. Ploughing and harrowing normally reduce the population of soil pests by exposing them to sunlight and desiccation, and to predators and parasites. Many Lepidoptera and Diptera feed on the aerial parts of plants as larvae but pupate in the soil. These soil-inhabiting pests would be depleted in numbers by normal cultivation methods but in areas where minimum cultivation techniques are employed the numbers often build up.

4 *Multiplication of suitable habitats* Farming leads to a simplification of the flora by a selection of plants suitable for husbandry. Thus pests associated with these plants have a more attractive and concentrated food supply. The most outstanding example of this is shown in the storage of grain and foodstuffs; many storage pests exist in small populations in the field but increase in numbers enormously in the favourable microclimate and abundant food of the grain store. Typical examples are to be seen in *Sitophilus* spp. on maize cobs (Fig. 1.4), *Sitotroga* spp. on sorghum, and bruchids on legume crops.

fig. 1.3 Caterpillars of *Plutella* sp. on cabbage

fig. 1.4 *Sitophilus zeamais* attacking stored maize seeds

5 *Loss of competing species* Under conditions of monoculture many insects which were not pests under natural conditions become pests, but the area has fewer insect species overall. Sometimes specific control measures may remove one pest, but another species released from competition pressure may increase in numbers and become a new pest.
6 *Change of host/parasite relationships* Most insect species are kept in check by their predators and parasites, although when a pest species increases in number there is typically a time-lag between its increase and that of the parasite/predator numbers. Parasites tend to be quite specific but predators less so, and the time-lag of the increase in parasite population is generally less than that of a predator. The greater the time-lag then the more likely is the species to be a serious pest. Generally agricultural operations involving large-scale applications of insecticide may affect parasites and predators more than the pests. One of the classical cases is that of the red spider mite (*Metatetranychus ulmi*) which has become a serious pest on fruit trees in many parts of the world after widespread use of DDT in orchards wiped out its predators. The giant looper caterpillar (*Ascotis selenaria*) was normally a minor pest on coffee in East Africa but it has become serious in places where the insecticide, parathion, has been regularly used over a long time.
7 *Spread of pests and crops by man* Pests may become established or uncontrolled when insects are introduced to countries where they did not previously exist. Classical cases are those of the gipsy moth (*Lymantria dispar*) in Canada, the Colorado beetle (*Leptinotarsa decemlineata*) in Europe, the Japanese beetle (*Popillia japonica*) and the hessian fly (*Mayetiola destructor*) in the USA, all of which caused very serious damage when introduced into new countries. Often in the new country the predators, parasites, and the more serious competitors for food are absent, hence allowing the population of the new pest to increase dramatically. When crop plants are introduced to new areas the indigenous insects may find the new food plants more acceptable than the previous ones, and many insects with polyphagous habits take advantage of such introductions.

Economic change A pest may arise purely for economic reasons, because of a change in the value of a crop. Damage that is not serious when prices are low can be very important when prices are high. The main economic reasons for a change in pest status are as follows.

1 *Change in demand* If some crops are replaced by others, then the pests of the former crops become less important. Greater demand may be for increased quantity and quality; both of these factors affect the importance of the pests. If the crop is in short supply then the consumers are less selective than if it is abundant. Wireworms do not greatly affect the yield of potatoes, but their tunnels spoil the appearance and keeping qualities of the potatoes. If the supply is short then the consumers will ignore a little damage, but with the increasing demand for packaged vegetables such potatoes are generally unsaleable and wireworms then become a more serious pest.
2 *Change in production costs* A pest may become economically important when agricultural practices change. For example, if a new high-yielding variety is developed, then minor pests which attack it may become of economic importance.

Pest damage to crop plants

There is an illustrated chapter on pest damage to crop plants in the companion text: *Volume 2 Field Handbook*.

Direct effects of insect feeding

Biting insects may damage plants as follows.

1 Leaves eaten, with subsequent reduction in assimilative tissue and hindrance of growth; examples are grasshoppers, caterpillars, sawfly larvae, leaf-cutting ants, leaf beetles and some weevils (Figs 1.5(a),(b),(c),(d) and (e)).
2 Leaves rolled and webbed, and eaten; examples are larvae of skippers, tortricids, and some pyralids (all Lepidoptera).
3 Leaves mined with either tunnel or blotch mines, by larvae of Agromyzidae (Diptera) or caterpillars of the families Lyonetidae and Gracillariidae, and some beetle larvae.

(a) *Dysdercus cardinalis* adults and nymphs on cotton

(c) Oil palm leaves damaged by locusts

(d) Larva of *Epilachna varivestis* on soya bean leaves

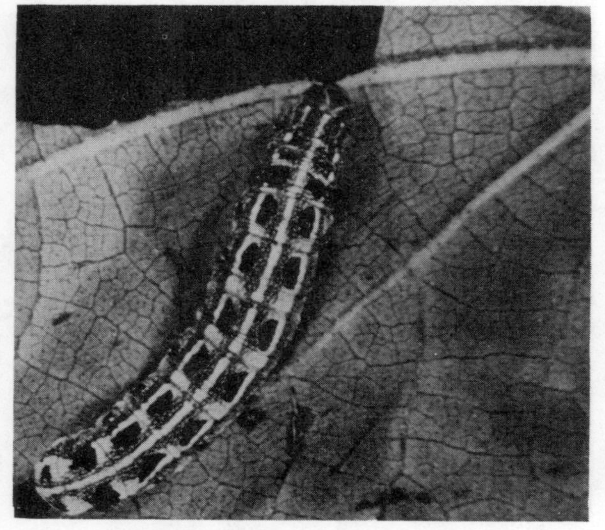

fig. 1.5 Insect damage to leaves (b) cotton leaf worm

(e) *Scyphophorus acupunctatus*, sisal weevil, adults feeding in sisal leaf exposing the peripheral fibres

4 Buds eaten, destroying either the growing point of young flowers and fruit, as by some grasshoppers, some beetles, tortricid larvae (budworms), and some other caterpillars.
5 Flowers and young fruit eaten, as by pollen beetles, blister beetles (Fig. 1.6), and some scarab beetles.

fig. 1.6 *Mylabris bicincta*, blister beetle on *Hibiscus* sp.

6 Fruits and seeds eaten or bored, and destroyed, as by sorghum midge larvae, pea pod borers, maize weevil, coffee berry borer, and various fruit flies (Figs 1.7(a),(b),(c) and (d)).

fig. 1.7 Insect pests of fruits (a) Maize weevil, *Sitophilus zeamais*

(b) Fruit flies of the family Tephritidae on peach

(c) Black maize beetle *Carpophilus lugukris* in USA

(d) *Lygaeus elegans* on sorghum in Botswana

7 Fruits bored and caused to fall prematurely, for example mango fruit fly, and coffee fruit fly.

8 Stems of both woody and herbaceous plants bored, with the subsequent death of the distal part of the stem, for example *Earias* spp. (Fig. 1.8(a)) in the cotton stem, *Diphya* spp. in coffee stems, larvae of Cossidae and *Sesamia* sp. (Fig. 1.8(b)).

9 Stems of seedlings bored, producing a 'deadheart'; for example *Atherigona* spp. larvae in cereal seedings, and *Chilo* spp. larvae in cereals.

10 Stem of woody plants ring-barked, as done by *Anthores* spp. on coffee, ,and *Indarbela* spp. larvae on cocoa, etc.

11 Roots eaten, causing a loss of water and nutrient absorbing tissue, as by chafer grubs (Coleoptera, Scarabaeidae (Fig. 1.9)), and some weevil larvae.

12 Tubers and corms bored, leading to a reduction of stored food material, and impairing both storage properties and next season's growth; examples are *Cylas* spp. weevils in sweet potato tubers (Fig. 1.10), yam beetles, and potato tuber moth larvae.

fig. 1.8 (a) *Earias insulana* larva on cotton stem

fig. 1.9 *Lepidiota mashona*, white grubs attack roots

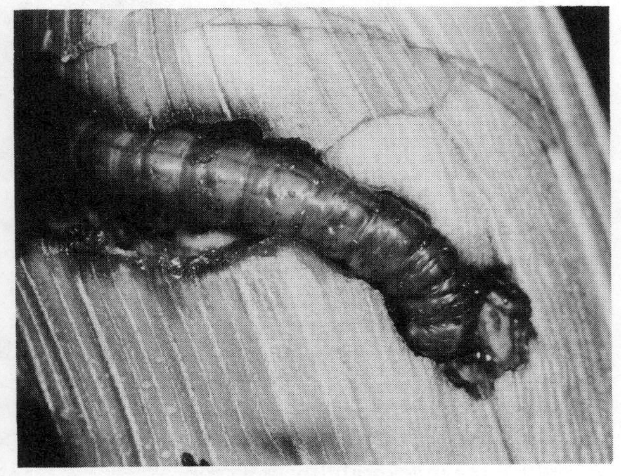

(b) *Sesamia* sp., stem borer on maize stem

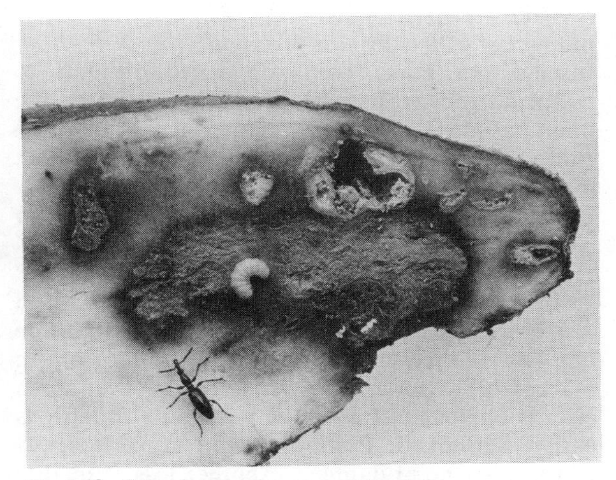

fig. 1.10 *Cylas formicarius* on sweet potato

Damage by insects with piercing and sucking mouthparts and mites.

1 Loss of plant vigour due to removal of excessive amounts of sap, resulting in extreme cases in wilting, followed by stunting of growth; for example most aphid species, and whiteflies on a range of crop plants.
2 Cause leaf-curling and deformation, as shown by aphids, thrips, mealybugs, white/blackflies and jassids.
3 Cause premature leaf-fall, as do many diaspidid scales.
4 Cause leaf and fruit scarification by rupturing epidermal cells and removing sap; as by spider mites and many thrips.
5 Toxic saliva injected by feeding bugs (Heteroptera) causes: premature fruit-fall (coconuts and *Pseudotherapterus*); abortion of young cotton bolls (*Calidea* spp.); death of floral parts and hence reduced seed production (coffee and coffee lygus bug (Fig. 1.11)); necrosis of cocoa pods (by cocoa capsids); stem necrosis and death of growing point (various mirid bugs, including cashew helopeltis); necrotic spots in young leaves resulting in leaf-tattering cotton lygus (Fig. 1.12)) and cotton leaves (Fig. 1.5(a) on p.9).
6 Provides physical entry points for pathogenic fungi and bacteria, as do *Dysdercus, Nezara* and *Calidea* spp. for the fungus *Nematospora* on cotton bolls (Fig. 1.13 on p. 13).

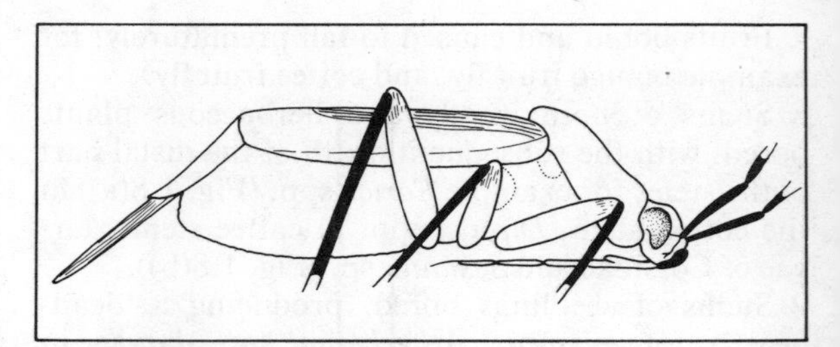
fig. 1.11 Coffee lygus bug

fig. 1.12 Cotton lygus bug

Indirect effects of insects on crops

Insects can make the crop more difficult to cultivate and/or harvest. They may distort the plant, as do *Earias* spp. larvae on cotton which cause the plant to develop a spreading habit which makes weeding and spraying more difficult. They may delay crop maturity, as do the bollworms on cotton. Grain in cereal crops may become dwarfed or distorted.

Insect infestation results in contamination and loss of quality in the crop. The quality loss may be due to reduction in nutritional value or in marketability (lowering of grade). Loss of yield in a crop is obvious but a loss of nutritional quality is easily overlooked. Pests of stored grain typically cause this type of damage. Another loss of quality is the effect of insects on the appearance of the crop, for example skeletonised or discoloured brassicas have a much lower market value than intact ones; skin blemishes, mealybugs and hard scales on citrus and other fruit reduce their value. Contamination by insect faeces, exuviae, and corpses; the black and sooty moulds which grow on the honeydew excreted by various homopterous bugs all reduce the marketability of a crop.

(a) *Dysdercus* sp. nymphs on cotton boll

Transmission of disease organisms Mechanical or passive transmission takes place through lesions in the cuticle caused by feeding. The pathogen, usually a fungus or bacterium, may be carried on the proboscis of the bug or on the body of a tunnelling insect. Examples are seen in the platygasterid wasp which can transmit coffee leaf rust, and *Dysdercus* spp. and *Nezara* spp. on cotton, in the latter case the spores are carried in the saliva of the bugs.

Most viruses are transmitted by an insect vector. The vector is usually also an intermediate host, as with most aphid and whitefly hosts. Common diseases transmitted in this manner include leaf curl of cotton (Fig. 1.14), cassava mosaic, tobacco mosaic and banana bunchy top.

fig. 1.13 Pests of cotton (b) *Nezara viridula*

fig. 1.14 Virus carried by white fly on cotton

2 Diseases

The diseased plant

Plant diseases can be defined in the widest sense as conditions of the plant involving abnormalities of growth or structure. It is this departure from the normal healthy condition, resulting in the appearance of disease symptoms, which enables diseases to be recognised. These can vary between death of the whole plant and minute changes in appearance only detectable to the trained observer. Frequently several symptoms, together or in succession, are produced by one disease, so that a symptom picture or **syndrome**, characteristic of a particular disease may be observed.

There are many factors which cause plants to appear unhealthy. In this book we are concerned only with diseases caused by pathogens. These are parasitic organisms which live in or on the host plant and cause the appearance of disease symptoms; the process is called **pathogenesis**.

Plants may also be damaged by pests (Chapter 1) and by mechanical forces, such as wind, hail and farm implements. There are non-parasitic diseases or disorders which are caused either by adverse environmental conditions or by internal physiological disturbances, usually of genetic origin. These include climatic damage due to frost, sun or lightning; mineral deficiencies or imbalances and genetic mutations. Parasitic diseases are important because they are infectious — they can spread between plants, often rapidly and extensively, and frequently produce epidemics.

The most important effect of disease for the farmer is the reduction in crop yield or quality which results. Unfortunately, the severity of symptoms when judged visually is not a good indication of the agricultural importance of the disease. Some diseases produce no immediately obvious symptoms in the field but they nevertheless cause substantial reduction in yield, e.g. ratoon stunting disease of sugar cane. By contrast, there are diseases which produce very obvious symptoms, such as the sooty moulds and some other leaf-inhabiting fungi, which have little effect on the productivity of the plant.

Symptoms of plant disease

Many diseases can be recognised immediately by the characteristic symptoms which they produce. Rust diseases are so-called because of their powdery, orange sporing pustules on leaves or stems, and smut diseases are recognised by the occurrence of black spore masses on the flowers or other parts of plants. More usually, symptoms of a certain type are produced by a variety of diseases. Brown spots on leaves are a common symptom and can be caused by a range of different parasitic organisms as well as by some pests and certain types of non-parasitic disorders. A further complication is that the same pathogenic agency can produce different symptoms on different hosts or even on the same host under different conditions. For example, *Xanthomonas malvacearum*, the cause of bacterial blight of cotton, can produce a seedling blight, angular leaf spots or stem lesions (Fig. 2.1).

fig. 2.1 Leaf spots produced by *Xanthomonas* sp.

Symptoms of plant disease can be divided into several groups. Localised symptoms, involve small parts of individual plant organs, such as leaf spots, or whole organs, such as root rots and blights. Systemic symptoms are those which affect the whole plant, such as wilts and general stunting. A similar type of symptom does not imply a similar distribution of the pathogenic organism throughout the plant. Localised leaf spotting may be caused by systemic pathogens, such as viruses. Others, particularly root diseases, cause systemic symptoms, such as wilting, even though they only occur locally in the roots. These systemic symptoms are secondary and are usually the first outward sign of root disease, the early stages of which cannot be seen in growing plants unless they are dug up. It is only when they become sufficiently severe to interfere with the general physiology of the plant that root diseases become apparent.

Diseases with systemic effects are usually more difficult to control in the growing crop and may cause more damage as they threaten the whole plant rather than a part of it.

Symptoms are usually described according to their appearance. Thus changes of foliage colour may involve chlorosis (yellowing), mottled patterns as in mosaics or leaf stripe disease, or the complete death of areas of tissues (necrosis). There may also be alterations in the growth form of the plant to produce dwarfing, or overgrowth of certain tissues as occurs in the production of galls or 'witches brooms' due to proliferation of lateral shoots. In addition diseases may produce many other less well-defined effects. Particularly with systemic diseases, many symptoms occur through secondary effects of the pathogen interfering with the normal physiological processes of the plant. Particular types of disease symptoms are described more fully in the *Field Handbook* which is a companion to this volume.

Control of plant diseases depends in the first instance on recognising when plants are diseased, and this requires an adequate familiarity with the healthy plant. The earlier a disease can be recognised the sooner can control measures be started and the smaller will be the losses caused.

Causes of plant disease

The measures used to control plant diseases are aimed, either directly or indirectly through the host, at the pathogen. Recognition of symptoms must be followed by identification of the causal organism, before the correct control measures can be applied. Some diseases such as certain rusts and smuts can be identified solely from their characteristic symptoms. However, in many cases symptoms can be misleading. As the type of control varies widely according to the characteristics of the pathogen, it is essential that the disease and its cause are correctly determined.

Parasitism

Parasites causing plant disease can be classed as either **obligate** or **facultative**, according to their dependence upon the host plant. Obligate parasites only grow directly upon the host plant, and cannot generally grow saprophytically on non-living organic matter. Their survival in the absence of a suitable host depends upon dormant resting stages in the life cycle, such as spores. However, facultative parasites are usually well adapted to a saprophytic existence and can survive long periods in an active stage in the absence of a suitable host. This difference in dependence upon the host plant underlies the choice of control measures

Obligate parasites depend critically upon the existence of the host. They cause only fairly mild symptoms, such as growth malformation, stunting and discoloration; they may not kill the host. Destruction of the host is of less consequence to the survival of facultative parasites which therefore cause more immediate and drastic damage, such as necrosis and wilting.

The intimate relationship between obligate parasites and their hosts is easily upset by genetic variations in either partner. These pathogens are generally very host-specific and different races of pathogen species are only able to parasitise limited groups of host species or varieties. For instance, there are 30 known races of the coffee rust pathogen *Hemileia vastatrix*. Each race is only able to attack a limited range of *Coffea* species or varieties within a species. Often the only practical way of controlling obligate parasites is by exploitation of host resistance. The breeding of plant varieties with various types of resistance to different groups of pathogens is a very complex subject, and will be dealt with in more detail in Part 2. Resistance to facultative parasites is usually less specific, partly because these pathogens have wide host ranges, but resistance is often affected by environmental conditions, e.g. charcoal root rot caused by *Macrophomina phaseolina* is most severe in plants affected by drought.

All pathogens must spend some time outside the host, in order to survive between cropping periods or to spread between plants. It is at this stage in their life histories that many control measures can be applied successfully. These include cultural methods and the application of protective fungicides.

Table 2.1 A simple taxonomic arrangement of the main orders of fungi containing important plant pathogenic fungi

Class Phycomycetes (Oomycetes)
Fairly primitive fungi with a simple sexual process and including many orders of simple water moulds.
Order
Peronosporales e.g. Downy mildews (*Peronospora* spp.), white rusts (*Albugo* spp.), blights, root rots and damping-off (*Pythium* spp. and *Phytophthora* spp.).

Class Ascomycetes
A large class of fungi characterised by the production of an ascus (sac containing typically 8 spores) during sexual reproduction and usually (apart from the yeasts) having specialised fruiting bodies (ascospores).
Order
Taphrinales e.g. Leaf curl (*Taphrina* spp.).
Eryisphales e.g. powdery mildews (*Erysiphe* spp., *Sphaerotheca* spp.).
Many other Ascomycete fungi in various Orders have parasitic asexual stages usually referred to as genera of the Fungi Imperfecti (see below). These fungi are generally unspecialised facultative parasites having saprophytic sexual (perfect) Ascomycete stages.

Class Basidiomycetes
Fungi producing a basidium (a cell producing spores from 4 protuberances) during the sexual phase, and often having complex fruit bodies.
Order
Ustilaginales e.g. smuts; Uredinales e.g. rusts; Agaricales e.g. many parasites of woody plants including Polyporaceae (bracket fungi), parasitic mushroom fungi (*Marasmius* spp. *Mycena* spp. *Armillaria* spp.) and some important parasitic genera with less obvious fruit bodies such as *Exobasidium* spp., *Corticium* spp.

Class Fungi imperfecti
Fungi without a sexual state and classified according to the asexual sporing characteristics. However, many are now known to be ascomycetes, but are still referred in the asexual state to genera of the Fungi Imperfecti. This class contains a very wide range of facultative parasitic fungi.
Order
Moniliales — producing spores freely on conidiophores and not enclosed in fruiting bodies, e.g. *Fusarium* spp., *Drechslera* spp., *Cercospora* spp., etc; Sphaeropsidales — spores produced in spherical fruit bodies e.g. *Phoma* spp., *Septoria* spp., *Diplodia* spp.; Melanconiales — spores borne in minute cushions or saucer-shaped acervuli, e.g. *Colletotrichum* spp.

Mode of action Parasites produce pathogenic symptoms in their host plants in a variety of ways. Obligate parasites live directly on living host cells and even this direct physical penetration of the plant tissues and utilisation of plant food may produce little immediate effect. However, steady debilitation of the host leads eventually to loss of yield. Facultative parasites may produce a variety of toxins and enzymes which kill the plant tissues as they are penetrated. Auxins which interfere with the growth-regulating mechanisms of the host are produced by both types of parasites. Many disease symptoms are the result of the defence reactions of the host, and include the production of gums and resins, the suberisation of cell walls, blocking of xylem vessels by tyloses, and the shedding of infected plant parts.

Fungi

The majority of plant diseases are caused by various parasitic fungi which are taxonomically diverse. Most parasitic fungi are facultative although some are specialised obligate parasites, such as the powdery mildews (*Erysiphaceae*) and rusts (*Uredinales*). Fungal pathogens, although diverse in form are characterised by the production of spores which enable them to spread between plants. Spores vary enormously in size, shape and their method of dispersal; these are important considerations in the control of plant disease (cf. p. 23). Table 2.1 lists the main groups of plant pathogenic fungi. No attempt will be made in this book to give an account of characteristics of the various taxonomic groups of fungi; these can be found in textbooks. Fig. 2.2 shows some fungi.

fig. 2.2 Examples of fungi parasitic on plants

(a) *Ustilago zeay*, smut on maize

(b) *Erysiphe graminis*, mildew on wheat (x70)

(c) *Puccinia* sp., rust on *Spiraea* leaf

Bacteria

Bacteria cause a number of important plant diseases, but many fewer than those caused by fungi. The genera *Pseudomonas* and *Xanthomonas* are by far the most important of the bacterial plant pathogens; they are relatively unspecialised. Although they are all facultative, some do exist in a number of races, *Pseudomonas solanacearum* for example. Each race is parasitic upon a different range of hosts.

Bacterial plant pathogens do not form spores but these single-celled organisms are motile and capable of limited movement in water films (Fig. 2.3). Spread of bacteria between plants occurs either by rain splash or carriage by insects, man and machinery. Since they can exist as saprophytes, they may persist in soil for relatively long periods, but they are usually killed quickly by desiccation.

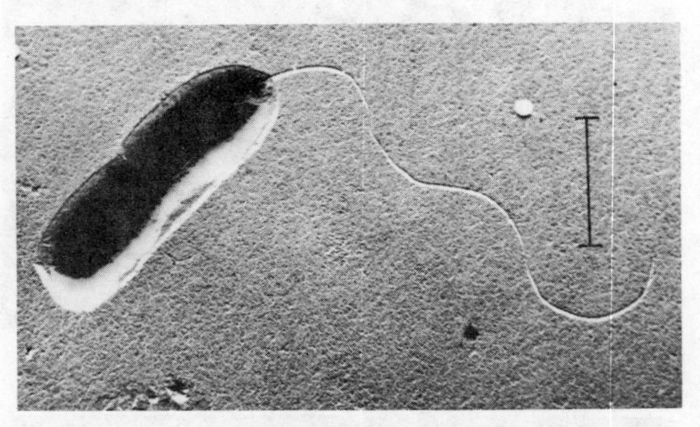

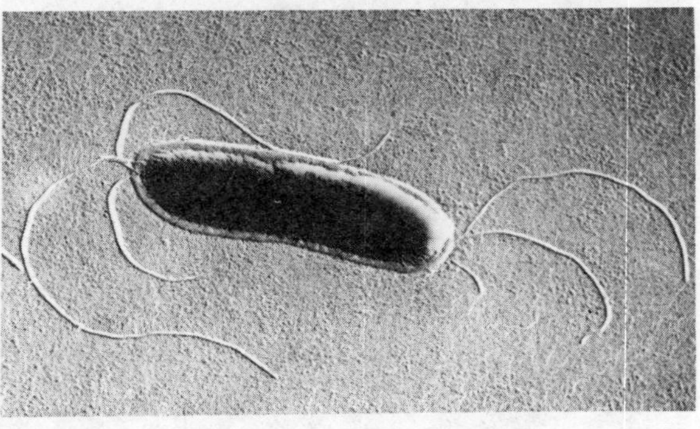

fig. 2.3 Plant pathogenic bacteria (*Pseudomonas* spp.) showing different flagella arrangements under the electron microscope. The scale in Fig. 2.3 (a) represents 1 μm.

Viruses

Plant diseases caused by viruses are among the most difficult to control. This is because knowledge of many plant viruses especially those infecting tropical crops, is still very limited. Furthermore, viruses are very difficult to work with and require elaborate and expensive equipment before they can be studied. Fig. 2.4 shows a selection of different types of viruses photographed under an electron microscope. They are very highly specialised obligate parasites and can only exist within living plant cells. Frequently they cause obscure symptoms, easily confused with mineral deficiencies or other environmental effects, and proof of the viral nature of a disease demands intensive research.

Most viruses spread between plants by means of living vectors, usually insects or nematodes, which themselves become infected with the virus, after feeding on a diseased plant. Some viruses can be transmitted by mechanical contact and others may be transmitted by fungi and in pollen.

Control measures against virus diseases are usually aimed at the vector, but resistant crop varieties and the use of clean planting material are also important. New virus diseases are constantly being discovered and recently another group of minute organisms called mycoplasmas have been shown to be plant pathogens.

Nematodes

Nematodes are minute non-segmented worms (roundworms), belonging to the phylum Nematode, which are common inhabitants of the soil. Some of these, usually of microscopic size, feed on or in the roots of plants and cause diseases such as root knots of many vegetable crops and root rots of perennial crops such as banana and citrus (Fig. 2.5a). They may also act as vectors of virus diseases.

Some nematodes move freely within the soil and feed externally on plant roots, whereas others may live inside the root tissue or remain permanently attached to the roots. It is usually young larvae of those nematodes which live in roots that move through the soil after hatching from their eggs, to find the roots of a suitable host plant. Sometimes the adult males remain free-living and only the females are parasitic within the roots (Fig. 2.5(b)).

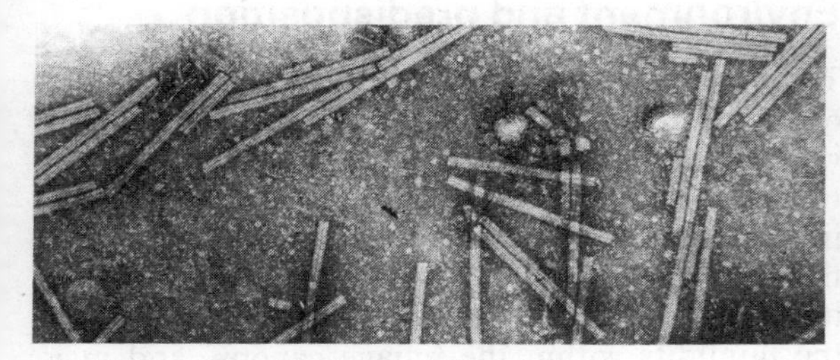

(a) Tobacco mosaic virus (x 100 000)

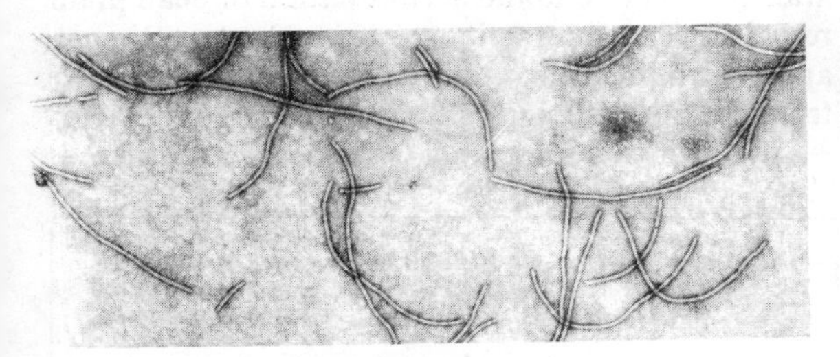

(b) Pepper veinal mottle virus (x 50 000)

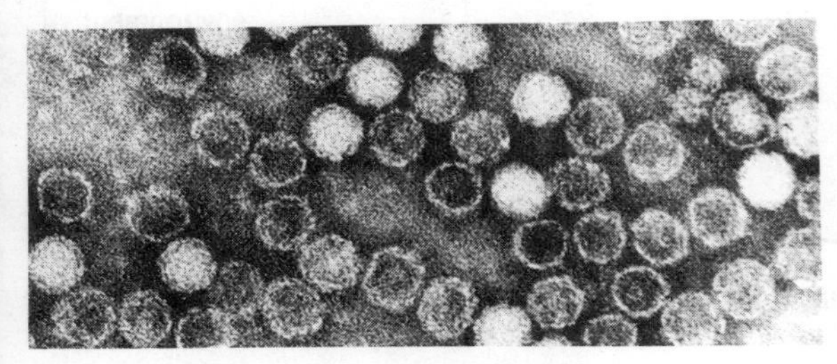

(c) Cocoa yellow mosaic virus (x 200 000)

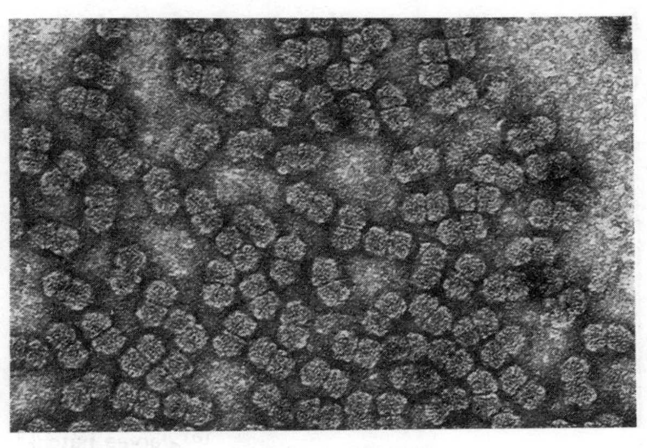

(d) Maize streak mottle virus (x 200 000)

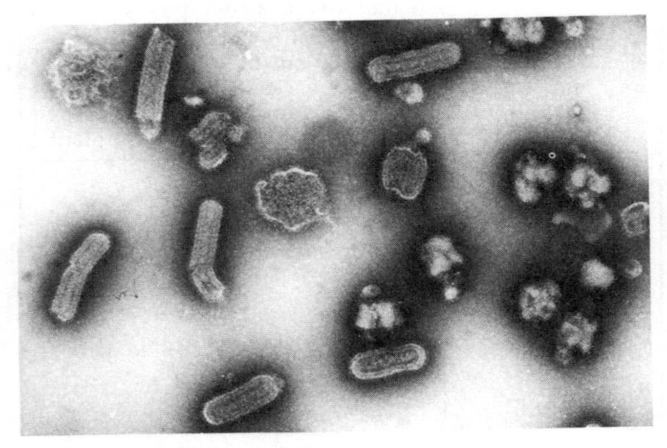

(e) Maize mosaic virus (x 100 000)

fig. 2.4 Electron micrographs of some plant virus particles

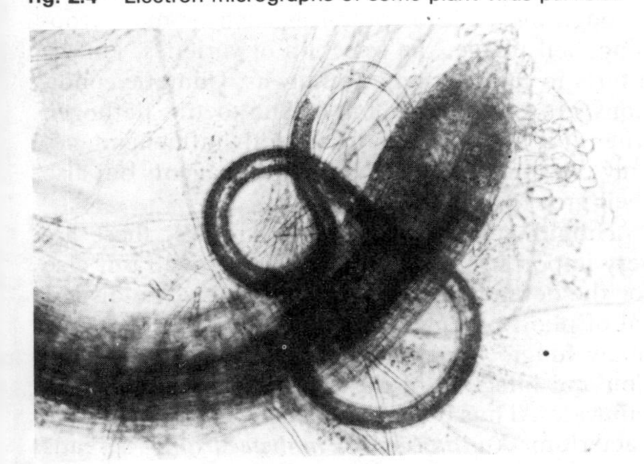

fig. 2.5 (a) Plant-parasitic nematode feeding on a root tip

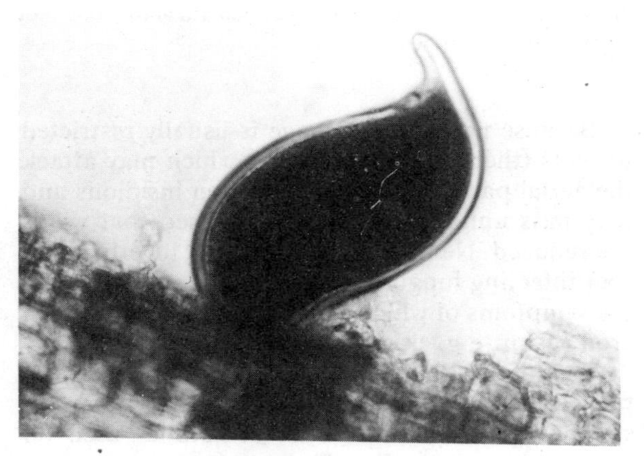

(b) Female parasitic nematode

Nematodes lay eggs either in cysts attached to roots, as with the potato cyst nematode, or free in the soil or roots of the host plant as with the root knot nematode. Fig. 2.6 shows a variety of life cycles of typical nematodes.

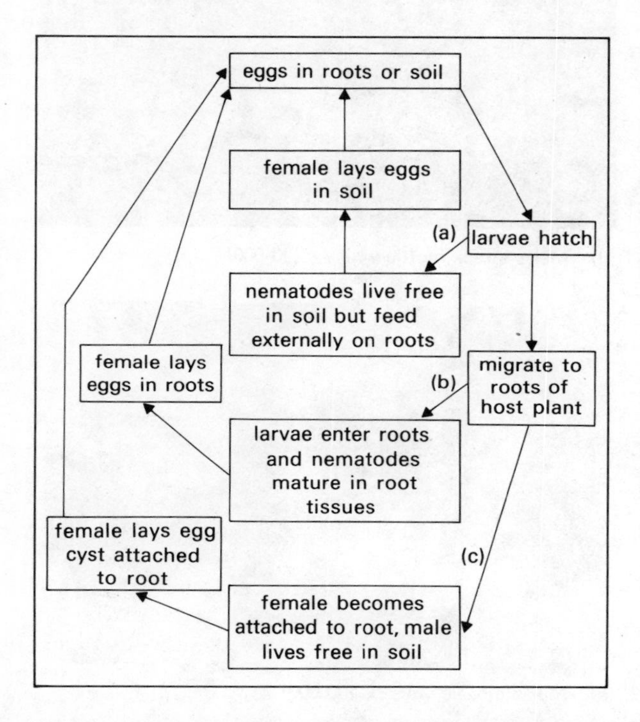

fig. 2.6 Life cycles of some typical plant-parasitic nematodes (a) free-living (e.g sting and dagger) nematode (b) root knot nematode (c) cyst nematode

Because nematode damage is usually restricted to roots (there are a few species which may attack the aerial parts of plants), it is often insidious and may pass unnoticed until it is realised that yields are reduced. Nematodes may also act together with root-infecting fungi to cause complex root diseases, the symptoms of which may be a general decline in crop vigour (e.g. citrus decline).

Nematode diseases are difficult to control and preventative measures are usually best, but certain soil fumigant techniques are sometimes used on perennial crops (see Part 3).

Environment and predisposition

The development and severity of plant diseases are often influenced by weather, soil and other environmental factors. Frequently these act by affecting the process whereby the pathogen reaches and infects the host plant, but they may also influence the susceptibility of a plant to a disease. The plant itself may also alter the environment by increasing the humidity within the foliage canopy, and may alter the soil due to the accumulation of dead plant remains. The interaction of host plant, pathogen and environment is often referred to as the disease triangle, but with agricultural crops a fourth factor — the influence of man — has to be considered.

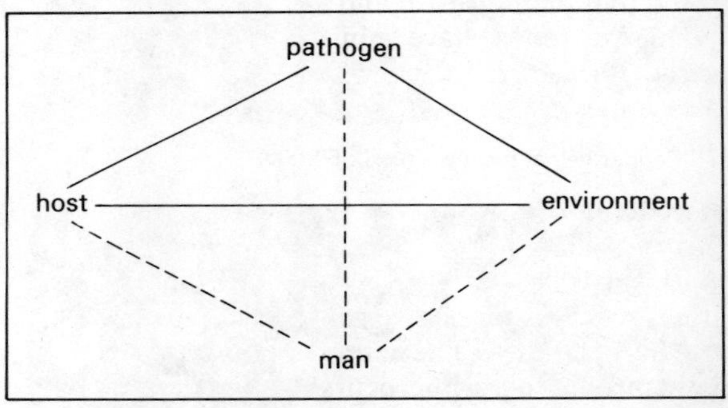

fig. 2.7 The disease triangle and square

This converts the disease triangle into a square as shown in Fig. 2.7. Man influences the relationship through agricultural practices such as monocropping, soil tillage, the selection of varieties, and by efforts to control plant pathogens. Climatic conditions are particularly important to the pathogen when it is outside the host, directly influencing not only the spread of many pathogenic agents but also their growth and survival.

Humidity and rain In the Tropics, these are very important, not only for plant growth but also for the development of plant diseases. The dispersal of plant pathogenic bacteria and the spores of many fungi, as well as their subsequent germination and infection of the host plant, depend upon rainwater. Thus black arm of cotton, caused by the bacterium *Xanthomonas malvacearum*, spreads most rapidly during driving rain, and black pod of cocoa, caused by the fungus *Phytophthora palmivora*, is most severe during wet weather. Often specific durations of wetness at certain temperatures are required for infection to take place after the spores have reached a new host. In view of the dependence of disease development on weather conditions, it is sometimes possible to forecast periods when diseases are likely to be severe. This enables control measures to be timed more effectively.

Seasonal variation of rainfall is the main factor influencing crop growth and plant disease epidemics in the Tropics. Appreciable daylength variations which influence crop growth in areas away from equatorial zones, and cold winters which halt plant growth and disease development in temperate zones, have minimal effects in tropical areas. Therefore, in the wetter equatorial Tropics, crop cycles frequently overlap and continuous unchecked epidemics of plant diseases often occur. Even where dry seasons do intervene the seasonal movement of the intertropical convergence zone and associated rainy seasons northwards and southwards across the equator carries pests and diseases across successive geographic zones.

Microclimate The climate is modified by the plant producing a microclimate within the foliage canopy. Because of the sheltering effect, conditions here may be very different from outside and are often favourable to plant pathogens because humidities are higher and exposure to temperature extremes is reduced.

Predisposition

Environmental factors can also influence the plant directly and can alter its resistance to disease by interfering with the functioning of its natural defence mechanisms. Diseases caused by facultative parasites may be of little consequence to the plant growing under optimal conditions, but can become severe when plants are predisposed to infection by stresses induced externally. For example, *Macrophomina phaseolina* causing charcoal root rot is more severe on drought-stressed plants. Physical or physiological damage caused by adverse environmental conditions or improper use of fertilisers can make plant tissues more susceptible to parasites. Although environmental predisposition is not an important factor in the development of many diseases, it should be taken into account in general agricultural practice, so that the impact of disease on production can be minimised.

The control of plant disease through altering the environment by cultural means is considered in more detail in Part 2. Agricultural practices such as monocropping, modification of the environment and selection of cultivars, as well as incidental spread, account for man's influence on plant disease. Economic considerations influence the relative importance of plant diseases.

Development of disease in the crop

Epidemics

The term **epidemic** is applied to the increase and spread of a disease, although in popular usage it is restricted to rapid and extensive outbreaks of disease. **Epidemiology** — the study of the increase and spread of disease — has helped considerably in the development of rational control measures against plant diseases. The term **endemic** is used to describe a disease which is generally prevalent throughout a particular area; an endemic disease may be also epidemic. Thus, coffee rust is endemic throughout most of Africa and Asia but seasonal epidemics of the disease occur in the rainy season.

Populations of all organisms have a natural tendency to increase and spread when conditions are favourable. This is responsible for the infectious nature of plant pathogens. The rate at which diseases increase depends upon the way in which they spread and which parts of the plant they parasitise. The availability and susceptibility of the host and the environmental factors, particularly climate, also affect the rate of increase. Some diseases, such as rusts and many other foliage diseases, increase rapidly during a single growing season; others, notably many root diseases of perennial crops, spread much more slowly. Plant pathogens often have well-developed reproductive and dispersal mechanisms to ensure the rapid colonisation of host plants or survival in their absence.

Epidemics start from an original source which

initiates the primary infection in a crop. This may occur in patches or 'disease foci', or may be more generally distributed if the source occurs throughout the cropping area. There are two kinds of spread:

1 where only primary infection from the original source occurs during the life of the crop and there is no secondary spread;
2 where primary infection is followed by successive generations of the pathogen which produce waves of secondary infection throughout the life of the crop. These spread very quickly and high levels of disease can result from a very small original source.

The **inoculum** is the quantity of a pathogen available for infection. This remains fairly constant during the course of diseases where primary infection predominates, but increases with successive generations of the pathogen in diseases characterised by much secondary infection.

Disease progress curves A disease progress curve shows how the proportion of diseased plants in a crop changes with time. Diseases where only primary infection occurs typically show a fairly constant rate of increase. Where secondary infection occurs the rate of disease development increases logarithmically until some limiting factor, such as the availability of healthy plants, causes the rate to fall off; this gives a typical sigmoid disease progress curve. Both types are illustrated in Fig. 2.8.

Disease progress curves provide a means of monitoring the progress of a disease and hence of deciding on the best time to apply control measures. Timing of control measures is important because adequate control at the beginning of an epidemic often results in very much less disease during the period of maximum growth of the crop. Delaying the onset of disease can therefore result in larger yields. In addition, the relative proportions of primary and secondary infection which occur during a cropping season affect the efficiency of different types of control measure.

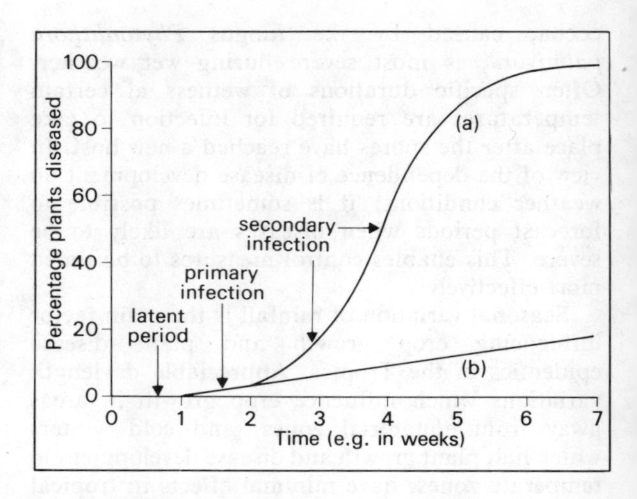

fig. 2.8 Disease progress curves (a) Rapid epidemic development with much secondary infection e.g. of a foliar disease (b) Slower but steady development with no secondary infection e.g. a soil-borne disease

Sources and inoculum potential

The sources of plant diseases can be divided into groups according to their location relative to the growing crop.

1 The soil and crop debris within it provide the source of many plant diseases. The soil is probably the most important source of facultative parasites, particularly those of annual crops. Usually these pathogens survive as dormant spores in the soil or grow slowly in the crop debris which is nearly always present in agricultural soils even when the host plants are absent; nematodes can survive as cysts or eggs. Virtually all diseases which attack roots and stem bases including damping-off of seedlings, and many foliage pathogens can be soil-borne. In soil-borne diseases, secondary spread may not be necessary because distribution of the inoculum is so widespread.
2 Crop plants themselves can be the main source of many disease outbreaks. Seeds and vegetative planting material can carry diseases in dormant stages; these develop as the plant grows and subsequently spread to other healthy plants in the crop. Bacterial blight of cotton, dry rot of maize and sugar cane mosaic are examples of diseases which can be initiated in this way. Vegetative planting material is an important source of soil-borne disease organisms, including nematodes. In perennial crops pathogens can exist in a dormant or

saprophytic state, in places such as bark, and initiate primary infection on susceptible stages of the crop. This occurs in coffee berry disease and black pod of cocoa.

3 Plants others than those in the crop, such as volunteers from previous crops and weeds, may act as sources of pathogens. These are important sources for obligate parasites such as viruses and mildews, and soil-borne diseases, including nematodes, which may survive on them in the absence of suitable crop hosts.

4 Disease may be initiated by spread from a distant source, often another crop of the same type. Efficient dispersal is essential for spread of diseases from distant sources. Obligate parasites of annual crops, e.g. rusts and viruses, may be carried long distances by wind-borne spores or insect vectors.

Inoculum potential Frequently the pathogen must be able to overcome a certain degree of natural host resistance before it can cause a successful infection, and the 'force' available to do this is called the inoculum potential. Where sources of the pathogen are very small or weak the inoculum potential may be too low to overcome host resistance, so that disease cannot be initiated. Similarly, if the initial inoculum is very scattered and present at a low density, the chances of its contacting a host may be very small. Reduction of initial inoculum, short of complete elimination, is thus often beneficial. Complete elimination of initial inoculum is only possible with seed-borne diseases, or where soil-borne pathogens can be killed by soil sterilisation.

Spread of diseases

The method by which diseases are spread, between individual plants, crops or seasons, is of special significance in epidemiology. Efficient dispersal is essential for the pathogen to reach new hosts and to survive from one season to another. Spread between plants may occur by direct vegetative growth; a method common with soil-borne pathogenic fungi, particularly those of perennial crops. *Armillaria mellea* and some other wood-rotting fungi can grow directly through the soil from the roots of one plant to another by rhizomorphs — root-like strands of fungal mycelium

fig. 2.9 Black root-like strands of mycelium produced by *Armillaria mellea*

(Fig. 2.9). This is fairly slow and occurs over short distances only. Nematodes may also move limited distances through soil to locate host roots.

Most pathogenic fungi are dispersed between plants by spores; these are small, specialised propagules, usually consisting of one or a few cells which can remain dormant but viable until they reach a new host. Fungi produce very many different spore types. Very small spores with thin walls cannot remain viable for long but travel fast and effect rapid dispersal. Larger spores with thick walls can remain dormant but viable for long periods, they enable the fungus to survive during periods when suitable host plants are absent.

Fungal spores may be dispersed in air currents or in rainwater, but secondary dispersal frequently occurs through carriage on objects such as farm implements or by animals. Fungi produce a wide range of specialised structures to facilitate spore dispersal — ranging from obvious fruiting bodies such as mushrooms to very simple rod-like conidiophores only visible under the microscope. Fungal diseases may be identified by these structures and by the spores themselves. Some types of spore and the structures which produce them are shown in Fig. 2.10 on p. 24.

Plant pathogenic bacteria do not produce true spores; they exist as single cells which are dispersed in rainwater or by secondary agents, which become contaminated with them.

(a) Cluster of conidiophores bearing long thin conidia of a *Cercospora* species on a leaf surface (× 250)

(b) Uredospores in a sorus of *Puccinia recondita* on a wheat leaf (× 50)

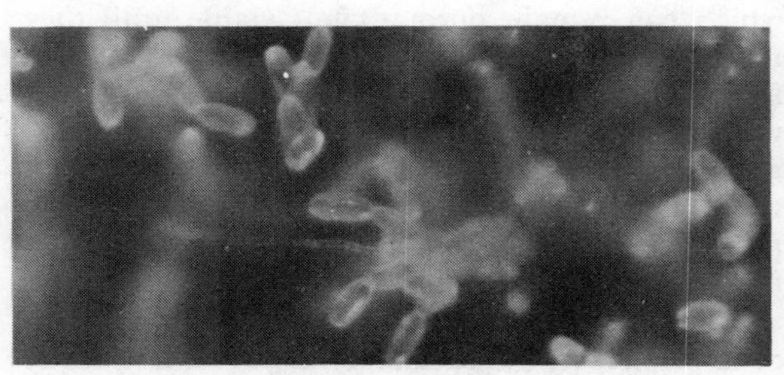

(c) Spores of *Erysiphe graminis* on a cereal leaf (× 250)

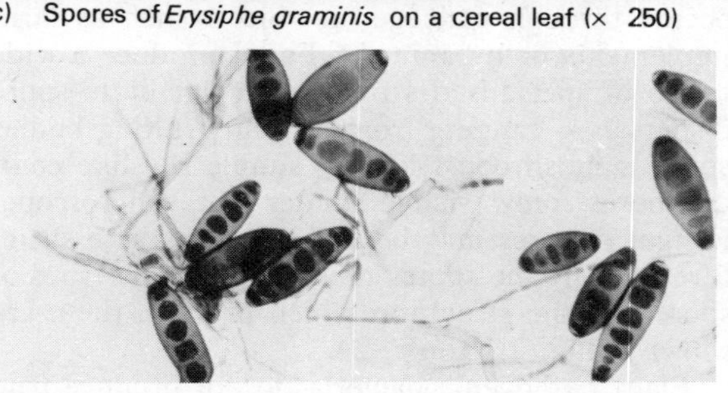

(d) Spores of *Cochliobolus sativus*, a cause of cereal leaf blotch and root rot (× 250)

(e) Long thin spores of a *Septoria* species squeezed out of the pycnidium in which they were produced (× 250)

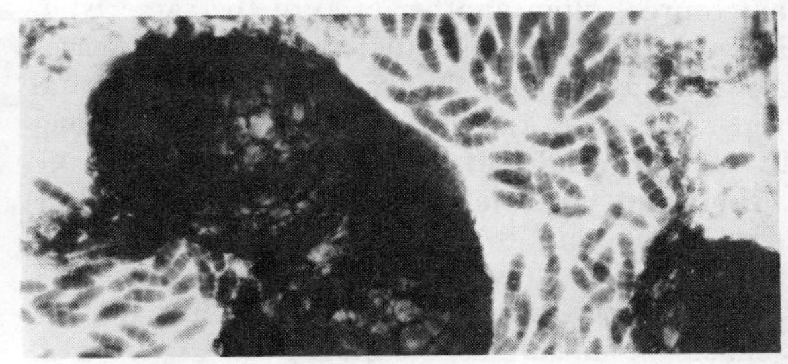

(f) Multicellular ascospores of *Pleospora herbarum* squeezed from the perithecia in which they were produced (× 250)

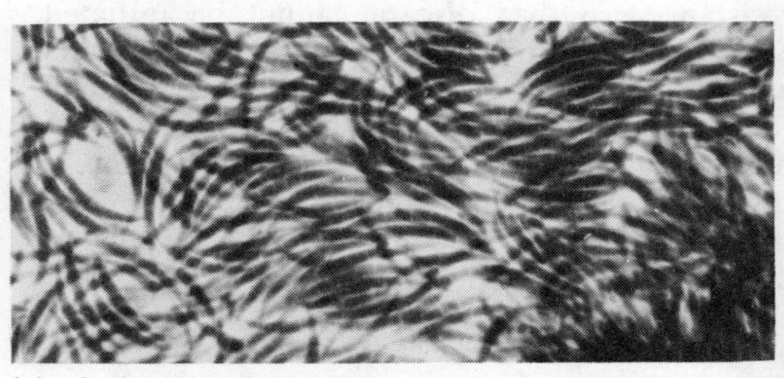

(g) Curved multicellular spores of a *Fusarium* species in a slimy mass (× 250)

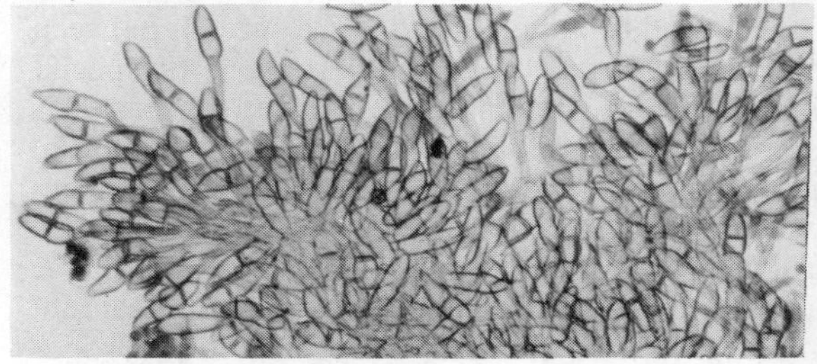

(h) Bicellular teleutospores of a rust fungus *Puccinia malvacearum* (× 200)

fig. 2.10 Examples of different spore forms produced by parastic fungi

Air Spores dispersed in air currents can travel very long distances if they become airborne and reach turbulent layers of air, e.g. 'clouds' of wheat rust spores, originating over Mexican wheat fields can travel northwards and infect Canadian wheat later in the season. In tropical areas, winds blowing towards the intertropical convergence zone carry spores of fungal pathogens across the equator. Airborne spore dispersal is characteristic of most obligate parasites, possibly because the chances of reaching a specific host are increased with wider dispersal.

Water Water-borne spores typically require rain for dispersal and are usually carried over much shorter distances. However, spores 'launched' by rain splash may travel 10 m from their source or may become directly airborne and be carried much further. Water-borne dispersal is limited to those times most favourable for spore survival and germination. Facultative parasites frequently produce water-borne spores since they do not require specific hosts.

Vectors Spores, bacterial cells or virus particles can be carried to healthy plants by vectors. These secondary agents are of special significance in the epidemiology of viral diseases, which rely on vectors for dispersal. Insects, particularly those with piercing and sucking mouth-parts, such as aphids, capsids, leaf hoppers and mealybugs, act as vectors for many viral diseases. Nematodes and fungi are now known to transmit many others.

Vectors acquire infective virus particles when feeding on or parasitising a diseased host. Insects transmit viruses in two ways:

1 non-persistent — the virus is transmitted immediately but only temporarily after feeding;
2 persistent — the vector becomes infective some time after feeding but remains so for a long period.

Control of a viral disease will depend on which way it is transmitted. Some vectors can remain infective for very long periods and the virus may survive between seasons in this way. However, survival on secondary hosts is more important.

Man Although most disease spread relies upon natural mechanisms and usually occurs over fairly short distances during an epidemic, dispersal through the movement of diseased plants by man is very important in extending the range of plant diseases. There are unfortunately many examples of disease which have been spread between countries in this way and the crossing of natural barriers such as oceans, mountain ranges and deserts, largely depends upon such transmission.

Some virus diseases can also be transmitted mechanically, especially by farm implements, which become contaminated with the sap of diseased plants; for example, tobacco mosaic virus.

Soil cultivation is also responsible for the distribution of soil-borne diseases including nematodes both within and between fields. The movement of man, machines and implements, seed and produce contributes to the rapid dispersal of diseases over large areas.

Infection

Infection of new host plants must occur after dispersal of the pathogen if spread of the disease is to be successful. This is a critical stage in the development of diseases. Of the many spores dispersed by pathogenic fungi very few reach suitable hosts and of those that do, very few succeed in establishing a progressive infection. This accounts for the enormous numbers of spores produced by these organisms to ensure survival.

Germination Spores must germinate before they can infect and this requires suitable environmental conditions, e.g. specific conditions of wetness or temperature or a suitable nutritional stimulus provided by a host root. Most soils are fungistatic to spores so that these remain dormant until suitably stimulated; this applies also to the cysts and eggs of nematodes.

Germ tubes produced by germinating spores either enter the host through natural openings, such as stomata or lenticels, or they penetrate the epidermis of the host by producing a specialised structure called an appressorium, which forces a fine infection thread through the host's epidermis (Fig. 2.11 on p. 26).

Some facultative parasites may grow over the host surface for some time before penetration occurs; this is very frequent among soil-borne pathogens. Obligate parasites must penetrate as

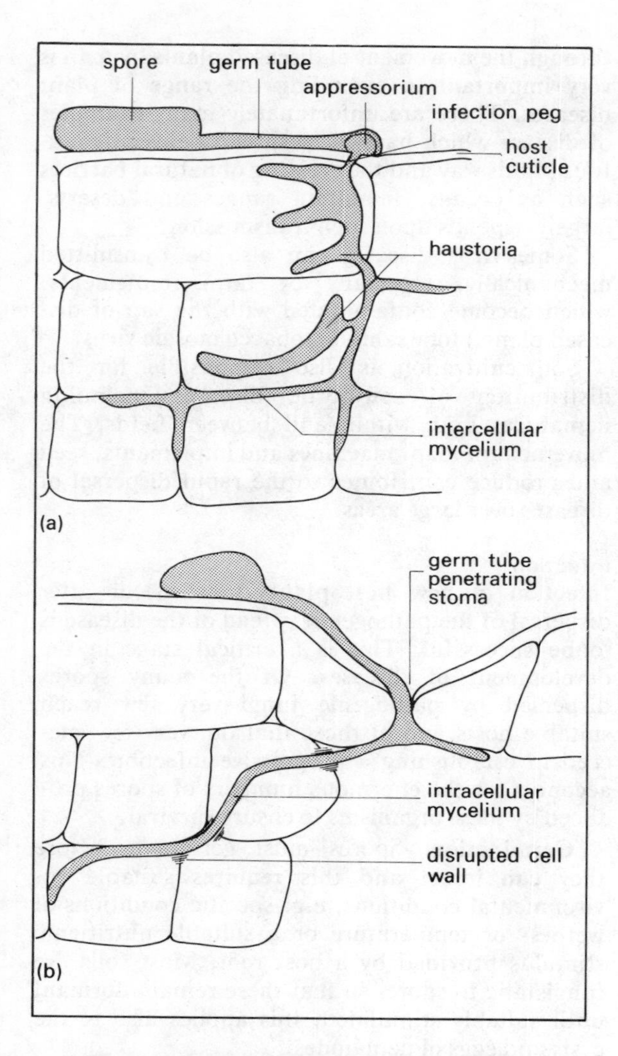

fig. 2.11 Fungal pathogens penetrating host (a) through appressorium showing intercellular hyphae with haustoria in the cells (b) through stoma, showing intracellar mycelium entering host cells by disrupting the cell wall with enzymes

soon as possible, as they cannot exist saprophytically, and most fungi attacking the aerial parts of plants cannot survive for long outside the host because adverse environmental conditions, such as high temperatures and desiccation, would rapidly cause their death. Many disease control measures operate just before the pathogen infects the host plant, when the pathogen is particularly vulnerable.

Bacteria usually enter through the plant's natural openings, e.g. stomata and lenticels, during very wet conditions.

Infection by virus particles occurs by their direct injection into the plant tissues by the vector. This is usually unaffected by environmental conditions except where virus particles are carried on the outside of mechanical vectors such as knives, where they may be rapidly inactivated by desiccation.

Growth After the pathogen has entered the host, successful disease development depends upon growth within the host tissues, this is only possible if the pathogen can overcome any host resistance. Hyphae of obligate fungi grow between host cells, and specialised structures called haustoria penetrate the cell wall, but not the cell membrane, and absorb nutrients. Hyphae of facultative parasites usually produce enzymes or toxins to kill the host cells and then enter by direct penetration.

Only certain parts of the plant may be susceptible; for example, most leaf diseases will not infect roots. Some diseases have very restricted zones of entry into the plant, e.g. smut diseases of flowers; some can only infect young plants, e.g. damping-off of seedlings; and others may only be able to infect growing tissues. Very many infections of plants which do occur are abortive because of successful defence reactions of the host. These may take the form of specific chemical substances that are toxic to the pathogen, such as complex chemicals called phytoalexins. Gums and tannins are produced by damaged cells to protect them from infection. Morphological barriers, such as thickened cell walls or corky scabs may also be produced.

Life cycle of the pathogen

The life cycles of all parasites involve growth in the host, reproduction and dispersal to new hosts. Viruses and bacteria have simple life histories that do not require the production of specialised fruiting structures. These pathogens can re-infect other hosts as soon as the number of viral particles or bacterial cells have multiplied sufficiently within the host to become available for dispersal by rain or vectors.

Fungi with simple life cycles only produce spores on the outside of the host for dissemination to other plants and this is a basic characteristic of the life cycle of all parasitic fungi. Before this can occur the fungus must grow vegetatively within the host plant. This phase is not usually infectious, except in some soil-borne pathogens, which can grow directly as vegetative mycelium through the soil between plant roots. Moreover, in many diseases caused by obligate parasites, symptoms of disease may not be apparent during this stage. This so-called latent phase of the disease is shown in Fig. 2.8 and affects the multiplication rate of the disease. Where it is short, as in mildews, which can start sporulating only a day or so after infection, the rate of disease progress is high, but in others the vegetative phase may last many months and the development of the epidemic is slower. In some diseases such as smuts, the life cycle of the pathogen is linked to that of the host, sporulation occurring when the plant flowers, infection of the developing seeds occurs, and the fungus begins to grow in the host plant when the seed germinates. The latent phase is important in disease control as during this period the plant is infected but not visibly diseased. This is why many disease control measures must be applied before the disease is apparent.

Just as many crop plants must survive unfavourable seasons in a dormant state so must the pathogens which infect them. Many pathogens survive in a saprophytic state on plant debris in the soil or on alternative hosts, but most fungal pathogens produce dormant resting structures, such as thick-walled spores or seed-like sclerotia, in response to the onset of unfavourable conditions. These germinate when conditions again become favourable, such as at the onset of the rainy season or when contacted by a suitable host.

During favourable conditions most fungal pathogens increase by the production of an asexual sporing stage which is readily dispersed to new host plants. When conditions are less favourable for rapid proliferation or when the host plant begins to senesce, the sexual phase often occurs. It is at this stage that genetic recombination allows the production of new pathogenic strains, although these may also arise at other times as well. The production of dormant spores, saprophytic states, or the ability to attack alternative hosts also occurs at this time so that the pathogen may survive between seasons during times when suitable hosts are absent. The various critical stages in the life cycles of plant pathogens are shown in Fig. 2.12. on p. 28.

The life cycles of many pathogenic fungi exhibit a complexity which must be taken into account when determining effective control measures. Fig. 2.12 shows the relationship between life cycle and control measures for a typical plant pathogen.

Different spores of the same fungus may be dispersed in different ways, they may survive for different periods and may infect different types of host plant. For instance, the rust fungi can have as many as 5 different spore forms, often produced in alternate hosts which may be necessary for the completion of their life cycle. Black rust of wheat, *Puccinia graminis*, for example, produces 2 types of spores on barberry, one of which infects wheat; another 2 types are produced on wheat the earliest of which (uredospore stage) causes epidemics of secondary infection, and a 5th type reinfects barberry (Fig. 2.13 on p. 29). Details of the various types of life histories of different rust fungi can be found in myclogical textbooks. Usually it is the uredospore stage of rusts which is responsible for the epidemic increase of disease such as occurs in maize rust, *Puccinia polysora*, and coffee rust, *Hemileia vastatrix*, but other spore stages produced from alternative hosts may be the major source from which epidemics of other rusts develop. The type of life cycle which different rusts have and whether or not alternate hosts are essential, affects the type of control measures which may be used.

Removal of the alternate hosts of cereal rusts has had little permanent effect on disease control as the uredospore form is able to carry the disease from one crop to another over long distances. Therefore, protection of the crop (usually by the use of genetically-determined resistance bred into varieties) is needed. It is, however, important to remove alternative hosts of soil-borne diseases such

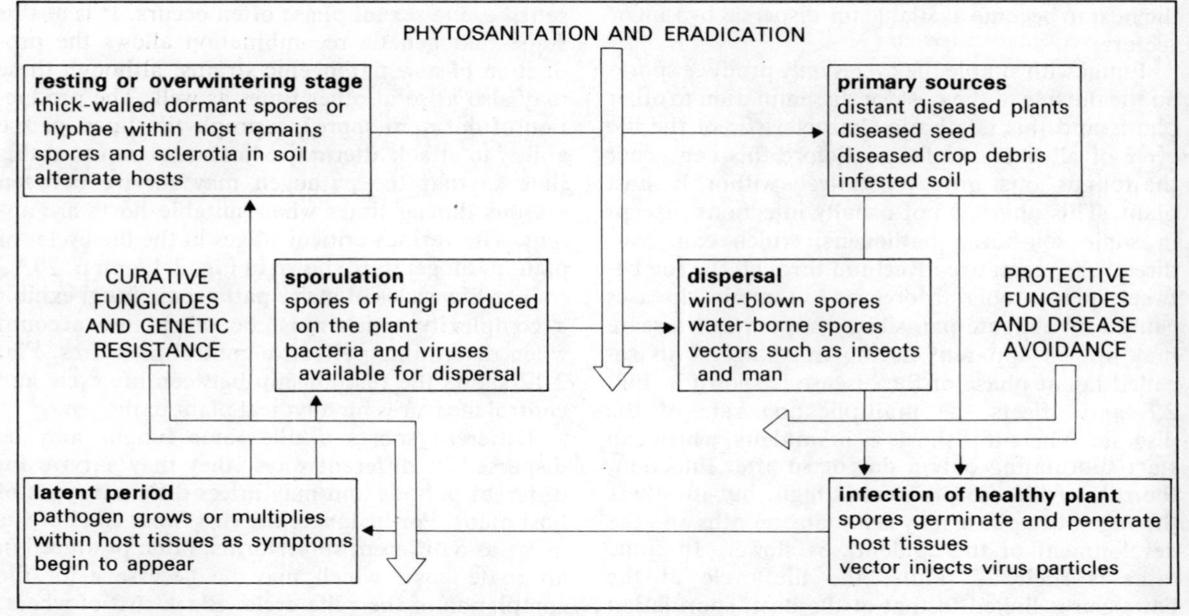

fig. 2.12 Life cycle of a plant pathogen showing effects of control measures

as bacterial wilt and root knot nematode as these provide important sources for survival during the absence of the crop host. Generally control of the widely dispersed airborne diseases such as powdery mildews, leaf spots of e.g. groundnut, tomato, depend on the use of host resistance or chemical protection. Whereas with soil-borne disease, crop rotation aimed at starving the pathogen from the soil is often very effective.

Epidemiology and control

Plant disease control in an agricultural context is concerned with preventing or at least restricting the development of plant disease epidemics; it is the whole population of crop plants which must be protected and the individual plant is of little consequence by itself. A knowledge of the mechanisms by which diseases develop and spread through crops is therefore essential before control can be rationally applied.

Most control measures for plant disease are designed to prevent rather than cure disease. They aim to operate on the pathogen before it has established a parasitic relationship with the host. The pathogen is more vulnerable and the chances of damaging the crop host by control measures are least when the two are apart. Elimination or reduction of the sources of initial inoculum are generally referred to as **sanitation**. These are often cultural practices and are of special significance for some diseases, particularly in preventing the establishment of those which are absent from certain areas.

Much control of plant diseases occurs after epidemics have started. These methods act during dispersal and re-infection and include the reduction of vectors, cultural practices aimed at producing a microclimate unfavourable for disease spread, and the use of protective chemicals.

The growth of the pathogen in the host can be prevented or restricted by the use of resistant cultivars and in some cases by the use of specialised chemical treatments. The stages at which plant disease control measures operate are shown above.

The removal of old diseased crop debris and volunteer plants helps to prevent the survival of many crop pathogens during periods when susceptible crops are not grown. Most common vegetable diseases can be controlled in this way. The use of

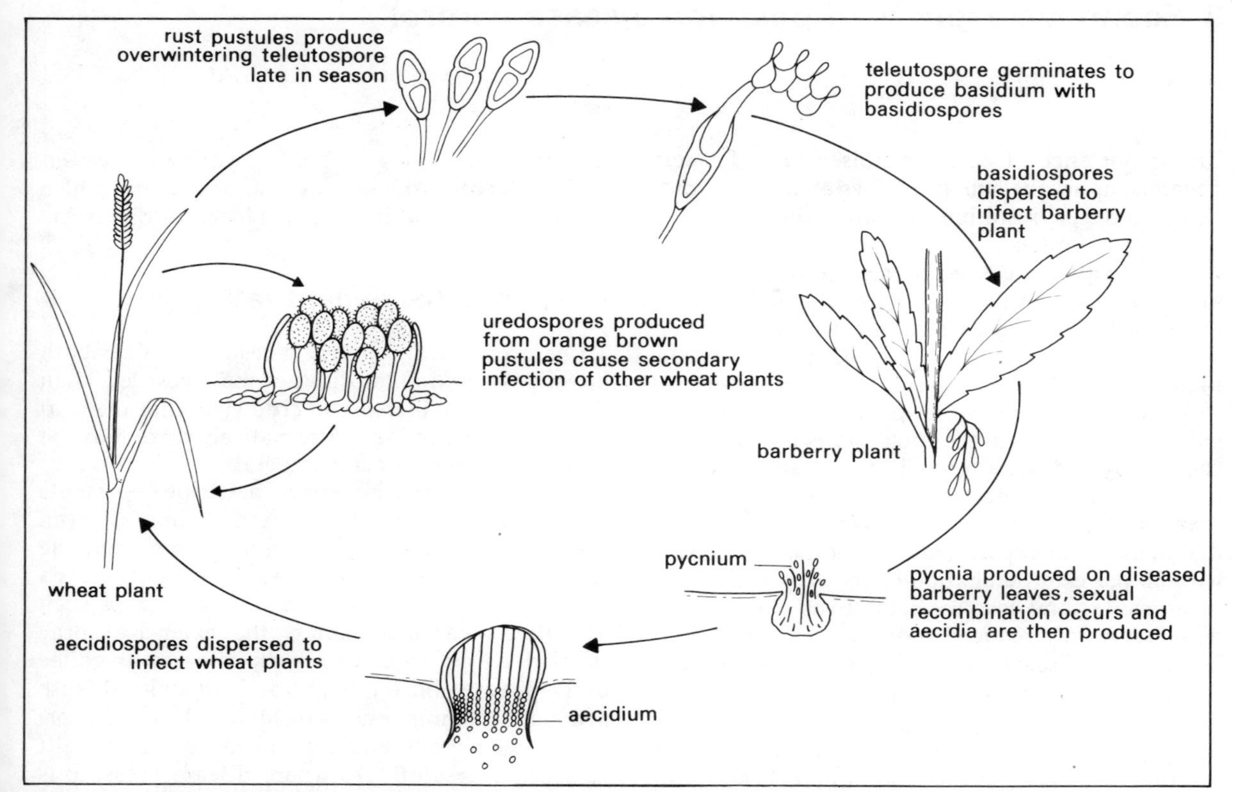

fig. 2.13 Diagrammatic life cycle of *Puccinia graminis* (black or stem rust of wheat) showing the various spore forms in association with wheat and barberry

disease-free seed or other planting material can also be regarded as a phytosanitary measure. The removal of diseased parts of perennial crop plants is also of special significance for control of some diseases, e.g. stripping of cocoa pods infected with black pod, *Phytophthora palmivora*. Legislative measures designed to prevent the entry of plant pests and pathogens into countries free from them is usually called **phytosanitary legislation**.

Further reading

Carter, W. (1973). *Insects in relation to plant diseases.* 2nd ed. Wiley: New York.

Commonwealth Mycological Institute (1981). *Plant pathologists pocketbook,* 2nd ed. Commonwealth Agricultural Bureaux: Slough, UK.

Dowson, W.J. (1957). *Plant diseases due to bacteria.* Cambridge University Press: Cambridge.

Jenkins W.R. and Taylor, D.P. (1967). *Plant nematology*. Rheinold: New York.

Large, E.C. (1948). *Advance of the fungi.* Cape: London.

Smith, K.M. (1974). *Plant viruses*. Chapman and Hall: London.

Tarr, S.A.J. (1972). *Principles of plant pathology*. Macmillan: London.

Wheeler, B.E.J. (1969). *An introduction to plant diseases.* Wiley: New York.

Wheeler, B.E.J. (1976). *Diseases in crops*. Studies in Biology Series, Edward Arnold: London.

Van der Plank, J.E. (1963). *Plant diseases: epidemics and control.* Academic Press: New York.

3 Economic aspects of pest and disease control

The importance of pests and diseases to farmers depends upon the amount of crop damage they cause and the resulting reduction in quality or quantity of produce.

The symptoms caused by disease or insect pest attack are by themselves an unreliable guide to the overall effect of a disease or insect pest on a crop. Total loss of or extensive damage to some of the plants in a field may have little effect on overall yield. This is because other plants in the field may gain from lack of competition and compensate for the loss by additional growth. For example, in the temperate zone, infestation of sugar beet by the mangold fly can lead to extensive and unsightly defoliation, but experimental work has shown that with 50 per cent defoliation of the crop there is no measurable total loss in yield. Even 70 per cent defoliation only reduced root yield by 5 per cent. Many virus diseases and some root diseases produce a general effect on all plants in a crop. The lower yield may then be considered normal and the disease may pass unnoticed. For example, with an infestation by early black bean aphid, bean yields may be reduced by as much as 30 per cent though the infestation may appear to be light. In some perennial crops any obvious loss in production due to a disease or pest attack may not occur until the year following the attack. This is the case with coffee rust, which reduces tree vigour and the production of bearing wood to carry the next season's crop.

The most effective way of assessing pest damage is to take many samples and to separate them quickly into different, easily distinguished, categories of infestation. To count numbers in a few samples is inefficient. A full assessment of the effect of plant diseases may require extensive trials to measure the quality and the quantity of yield from crops with varying levels of disease at different times during the life of the crop. The measurement of disease presents special problems and is usually based on the severity of symptoms for each disease, e.g. degree of wilting or percentage leaf area attacked by leaf spots (for which visual keys have been published for some diseases).

Cost/potential benefit ratio

To be economical, control measures against both insect pests and plant diseases must cost less than the value of the increase in crop yield that the control measures produce. Alternatively there must be some other long-term advantage.

Thus, the extensive use of pesticides on staple food crops is not generally practised because even though yields may be substantially increased, the value of the extra produce does not offset the costs involved. With high-value export cash crops such as coffee, tea or bananas, the extensive spray schedules used to control such diseases as coffee berry disease, blister blight and Sigatoka disease produce economic returns on all but the most poorly managed crops and may produce cost/benefit ratios in excess of 1:10 where disease pressure is high.

The potential benefit from controlling insect pests and diseases of plants must depend on the magnitude of the losses caused by the absence of control, and the efficiency of the control measures. These factors vary from season to season and control measures which are economic in one year may be uneconomic the next. In some instances, although disease incidence may be high, other features such as the relationship between the size of the crop and the price, may limit the usefulness of any additional yield obtained by disease control.

There is, therefore, a high degree of uncertainty about the economics of any specific control measure. This is particularly so in the control of plant disease where the best results are normally obtained by the application of control measures before the symptoms of disease become obvious. For example, the application of a fungicide spray just before a period of infection can achieve much better control than several sprays applied too late. It is, therefore, of great importance to forecast disease and pest infestation where possible. This consideration has not received much attention to date in developing countries, see further discussion on p. 34 et seq.

In practice, cost/potential benefit ratios are only known for high-value crops such as citrus and coffee. When they are available they tend to apply only to the benefit likely to result from control of the heaviest infections. When the ratio is high, uncertainty as to the likelihood of control can to some extent be ignored. Seed dressings are usually a cheap form of protection and when used to control insect pests such as the cereal shoot fly, for example, can provide cost/potential benefit ratios as high as 1:10. A ratio of 1:10 means that there only has to be an attack by the pest once in 10 years for the control measures to be justified in economic terms.

The national economics of pest and disease control programmes

A major proportion of the cost of pest and disease control programmes applied by the farmer is reflected in the retail price of food and is ultimately borne by the consumer. On the other hand the cost of quarantine, of the eradication of imported pests and diseases, of biological control and the major cost of the research required to obtain knowledge on the control of pests and diseases are usually provided by the government and funded from tax revenue.

The national benefit of pest and plant control programmes may be very different from their apparent economic value at the field level. The use of chemicals on a vast scale has undoubtedly had a profound and often deleterious effect on the total environment. In practice, the overall economic value of the control measures utilised may often be negative. Thus, it must be emphasised that pest and plant disease control measures cannot be considered in isolation. Their overall long-term effects on the environment and the social cost must be evaluated.

Investment in adequate plant quarantine facilities and in the eradication of exotic pests and diseases which may be accidentally imported may be very economic measures. For example, after the Mediterranean fruit fly had been accidentally introduced into Florida, a total of 0·7 million hectares of citrus trees and adjoining land was sprayed at a total cost of $7–9 million. This operation did save Florida's major agricultural industry, however, and the total cost was no more than about 5 per cent of the annual gross value of the citrus crop (Figs 3.1 and 3.2).

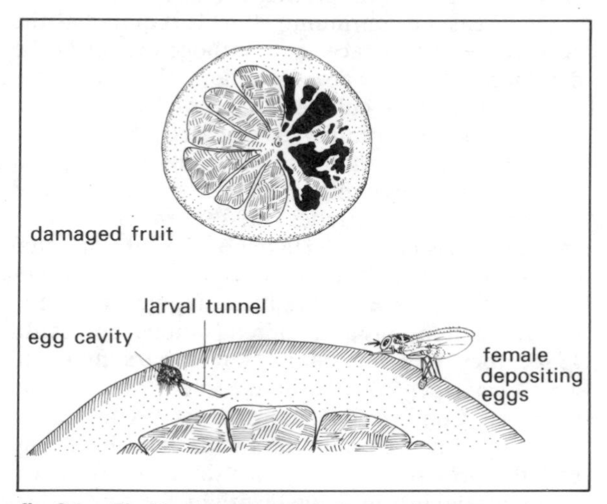

fig. 3.1 The Mediterranean fruit fly *Ceratitis capitata* deposits eggs through the skin of the citrus fruit; the larvae tunnel into the fruit causing extensive damage

fig. 3.2 Adult Mediterranean fruit fly, *Ceratitis capitata*

The cost of biological control programmes is very difficult to estimate but the Department of Biological Control of the University of California has stated that for the period 1923 to 1959 a total expenditure on biological control of $3.6 million resulted in a saving of about $100 million in California. Generally the costs of biological control compare very favourably with the costs of chemical control.

The development of resistant plants to control plant diseases receives government financial support in most countries. Sometimes this has achieved outstanding and permanent success, but for many diseases a continuing effort is required as the occurrence of new races of a pathogen requires the development of new crop varieties. Also the potential of many new cultivars has not been realised because of susceptibility to diseases; this may be regarded as an indirect loss caused by plant diseases. The breeding of new high-yielding and disease-resistant varieties of the world's major staple crops is now undertaken by the International Centres for Agricultural Research. These are funded by international donors including governments, private foundations and United Nations' bodies. In 1976, they received 64 million dollars much of which was spent on the development of new crop varieties. Commercial enterprise is becoming increasingly involved nowadays with plant breeding, and the provision of pure, healthy seed and planting material is largely dealt with by the private sector in many countries.

Disease surveillance and assessment

When assessing the general health of a crop it is necessary to relate the appearance of the crop to such factors as time and rate of sowing, variety of crop, soil and weather conditions. Diseases usually become evident when symptoms are very noticeable and unfortunately this is often too late to undertake effective control measures. Nevertheless, the recording of the incidence of diseases is necessary in order to determine how much damage they cause and how their prevalence is related to other factors. Often diseases commence as discrete patches or foci within a crop which are best observed by viewing from above. Aerial photographs are very useful in locating disease foci in areas under extensive monoculture that cannot easily be scanned on foot. However, there is no substitute for walking through crops and looking out for disease symptoms such as chlorosis, wilting, die-back or discrete lesions on leaves or fruit. Should these be found it is necessary to send samples to a competent pathologist to diagnose the causal organism.

fig. 3.3 A leaf of a banana plant showing symptoms of Sigatoka leaf disease

It is often necessary to measure the amount of disease present in a crop to assess whether it is sufficiently prevalent to cause economic damage and thus require control. For some diseases, the relationship between the amount of disease at a certain stage in the life of the crop and the reduction in yield it causes has been established. This is known as a disease/loss relationship and has been worked out for a number of diseases of tropical crops such as Sigatoka disease of bananas (Fig. 3.3) and some cereal leaf pathogens. It is also desirable that measurements of disease prevalence are made so that fluctuations between seasons, cultural treatments, crop varieties can be analysed and used to improve control measures.

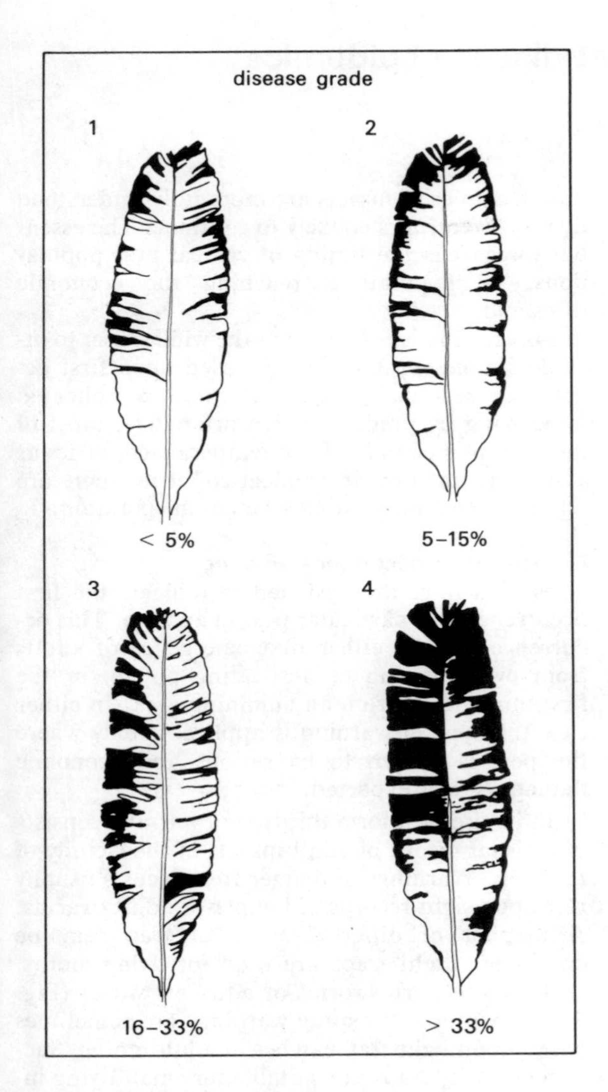

fig. 3.4 Banana leaf spot assessment key

The relationship between the amount of disease and the loss in yield or quality of produce that it produces is often very complicated. The simplest relationships are where the disease directly destroys the harvestable produce, as in fruit-rotting organisms such as coffee berry disease and grain-destroying organisms such as smuts. When the disease kills whole plants, an early attack which kills young plants and leaves gaps in the crop may be compensated for by greater growth of the healthy neighbour, on the other hand later attacks on older plants may occur after these have already produced some yield. With foliage diseases, early attacks generally have a great effect on subsequent growth and yield of the crop than do later attacks.

Basically, there are two ways of measuring disease. Firstly the **incidence** of a disease is recorded as the proportion or percentage of plants in a sample which have symptoms. Secondly, the **severity** of a disease is a measurement of the proportion of the plant affected by the disease, thus a leaf spot may affect a certain percentage of the leaf area or a die-back disease may affect a certain proportion of the branches on a tree. Because diseases tend to increase during the life of a crop and the effect that they have on yield also varies with time, it is important that diseases are assessed during a critical time in the life of the crop plant. For several important crops such as wheat, rice, soya beans, recognised growth stages can be readily identified from visual keys, and for many important leaf diseases of those crops, visual keys have been drawn up to enable an easy assessment of the severity of the disease. These are based on outline drawings depicting the percentage leaf area affected at certain intervals of severity (Fig. 3.4). The crops for which disease assessment methods have been worked out are listed in the FAO/CAB Crop Loss Assessment Manual, and most of these depend upon assessing both the incidence and severity of diseases in crop samples at critical growth stages.

Further reading

FAO/CAB (1970-). *Crop loss assessment methods*. FAO: Rome.

4 Forecasting pest attack and plant disease epidemics

Forecasting pest attack

Pest control programmes aim at accurate forecasting of pest attacks so that control measures can be planned in advance with maximum efficiency. Successful forecasting techniques should be based upon detailed knowledge of the biology and ecology of the pests concerned; they should be kept as simple as possible. The types of detailed studies required to give the basic information required for accurate pest forecasting are as follows.

1 **Quantitative seasonal studies** These must be made over at least several years to determine the limits of seasonal occurrence, variability in numbers, and geographical distribution. These studies must use sampling methods appropriate to the pest and its abundance. All seasonal counts and population estimates should be related to weather and topographical data.

2 **Life history studies** These are carried out to find the number of eggs laid, length of time required for each stage of development, amount of food eaten by each instar, maturation period for the females, etc., and are aspects that can be studied both in the field and in the laboratory. Behaviour of different larval instars and the possible number of generations under different conditions can be most suitably studied in the laboratory. The expected range in ambient temperature and relative humidity in any area can be considered in relation to the limits of survival of the pest under study.

3 **Effects of weather on the pest** Either directly or indirectly climatic factors control pest numbers, affecting not only the host plants and the pests themselves but also their predators and parasites. The spread of pests from crop to crop is always influenced by weather. Field studies are required to understand the effects of weather on the pest.

Pest forecasting

Accurate pest forecasting is very difficult and the many forecasting schemes in different countries have met with variable success. Until the population dynamics of insects are more fully understood this state of affairs is likely to continue. The essential forecast is the timing of critical pest populations, or populations reaching the economic threshold.

Forecasting is used here in the wide sense, to include simple spray warnings based upon first occurrence records, as well as more complicated forecasting by prediction. Unfortunately, most of the examples cited are from temperate countries as at the present time in tropical countries there are not many pest-forecasting schemes in operation.

Emergence or occurrence warnings

These warnings are designed to indicate the first occurrence of a particular pest in an area. This occurrence may be either first emergence of adults from overwintering or aestivating pupae, or the first immigrants from an adjoining area. In either case this type of warning is applicable only where the pest is known to be serious and economic damage is to be expected.

In its simplest form this type of warning consists of a visual record of adult insects in the vicinity of the crop. Warnings of danger from locusts usually rely upon sight records of hopper or adult swarms. A number of different types of trap may be employed. Light traps are used for flying moths, such as adult armyworms or adult cutworms (Fig. 4.1). Codling moth spray warnings are sometimes based upon light-trap catches of adult moths. Suction or sticky traps are suitable for small flying insects like aphids, leafhoppers and midges. For similar small insects simple water traps can be employed — these are used for catching adult gall midges in cereal crops in Scandinavia and cabbage root flies in the UK. Emergence traps are most suitable for insects which pupate in soil or litter, such as Diptera and Lepidoptera. Notices of spray warnings have to make due allowance for the female maturation period, rate of oviposition and time required for the eggs to hatch under the prevailing weather conditions.

fig. 4.1 Light trap for insect pests in cotton

Insect monitoring Pheromones (p. 00) are ecto-hormones secreted in the Lepidoptera (and other insects) by virgin females, to attract males for the purpose of mating; they are effective for long distances downwind. The components of the pheromones of many crop pest moths have been identified since about 1960, and in recent years many have been synthesised and are now available commercially. Pheromone-baited traps can be used to detect the presence of a pest, and also to estimate its population density (Fig. 4.2). Thus, conventional control measures can be precisely timed for maximum effectiveness and employed only as required. This procedure is known as insect monitoring and is used widely for moth pests in fruit orchards in South Africa, North America and Europe, and for cotton crops. It has been successful in monitoring fruit fly populations in citrus orchards.

Attractant traps can also be used for direct population suppression. Here control is achieved when the traps are employed at a sufficient density in the crop area to reduce mating success.

Forecasting by sampling

By sampling immature stages of insect pests it is possible to arrive at an approximate estimation of numbers of adults to be expected later. One of the more reliable methods of pest sampling in this manner is by taking soil cores for insects eggs (cabbage root fly, carrot fly), or beetle larvae

fig. 4.2 Pheromone trap for *Heliothis* spp., a pest of cotton

(wireworms, chafer grubs), or pupae of Lepidoptera and Diptera. With pests that have alternative hosts the pests may be sampled while on the other host so that an estimate of the pest density on the crop can be made.

The use of sampling results to predict the possible numbers of the damaging stage of an insect pest is clearly quite difficult. Years of observations and collection of empirical data are usually necessary before forecasting can be attempted successfully. For example, egg mortality can be very high. At Wellesbourne it was recorded that 80 per cent of cabbage root fly eggs were eaten by soil predators, and in Southeast Asia the proportion of rice stalk borer eggs destroyed by natural enemies was in the order of 95 per cent. Thus the sampling of fly eggs in the soil to give an estimate of larvae later in the season is not straightforward, and the sampling of fly pupae for the same purpose presents many problems. However, for some pests sufficient data have been accumulated for a reasonably accurate system of sampling and prediction to be used successfully.

In the case of Lepidopterous pests the best spraying date often may be determined by the occurrence of the first eggs on the crop. This is the method at present employed with pea moth on pea crops, and bollworms (*Diparopsis* spp. and *Heliothis* spp.) on cotton in Malawi and the Sudan. When the first eggs are found on the foliage of sample plants from crops in areas at risk, allowance is made for egg development, and the spray dates determined so that the first instar larvae can be destroyed (Fig. 4.3 (a) and (b)).

(a) Eggs of *Busseola fusca* under maize leaf sheath

(b) Eggs and adults of white fly *Bemisia* sp. on cotton leaf

fig. 4.3 Insect eggs on foliage

Forecasting by prediction

Temperature is probably the single most important environmental factor controlling the rate of insect development and hence population numbers. A simple method of using mean temperatures for 2 months has been developed to predict the date of emergence of the adult of the rice stem borer (*Chilo simplex*) in Japan. It has also been used in the USA for the prediction of outbreaks of European corn borer (*Ostrinia nubilalis*). The development of the pest is indicated by the number of degrees above 10 °C each daily mean temperature reaches. The accumulation of these departures from the base temperature in one season are expressed as degree-days. Observations over several years have established the relationship between degree-days and the different stages of development of the insect pest.

Rainfall recordings have also been used to forecast the likelihood of pest attack. In Tanzania outbreaks of the red locust (*Nomadacris septemfasciata*) have been successfully forecast from an index of the previous year's rainfall, and in the Sudan the amount of pre-sowing rains has frequently enabled jassid damage to the cotton crops to be predicted.

Some areas where critical infestations of some pests are likely to occur can be predicted from observations of climatic areas. The principal factors controlling a build-up of a pest population may be climatic, biotic or topographical; a combination of temperature and relative humidity or rainfall is probably the most important.

Geographical distribution of many pests is controlled by some limiting climatic factor. The distribution of insect pests can be divided roughly into three geographical regions or zones.

1 *Region of natural abundance* — here the insect is always present in detectable numbers and is regularly of pest status.

2 *Region of occasional abundance* — here the insect population is kept at a low level by climatic conditions and can rise to pest proportion only occasionally.

3 *Region of possible abundance* — sometimes, but rarely, the climate permits a population outbreak to occur. Insects often move into this region from

regions 1 and 2 and may be a pest for a while until the climate controls them.

Forecasting plant disease epidemics

Disease control measures are mostly preventive rather than curative in action. Prediction is therefore essential so that adequate control can be applied. For example, many fungicides are **protectants**, i.e. they are applied early to protect against infection. In most cases, past experience is a reliable guide and diseases which have been troublesome in the past should be guarded against. However, the incidence of disease epidemics may vary from season to season and it is useful to forecast when diseases are likely to become severe. A knowledge of when infection is likely to occur may improve control and may reduce the need for repeated application of fungicide. A well-timed fungicidal spray, applied just before infection, will achieve better control than several applied too late.

To forecast a disease the presence of the pathogen must be established first. The presence of common airborne pathogens, e.g. rusts, mildews, and other leaf diseases of many crops, can usually be taken for granted as they are easily spread from distant sources. With the rapidly developing airborne diseases, it may be too late to wait for the first appearance of the disease in the crop. The occurrence of airborne diseases can sometimes be predicted if they occur nearby or if they have been detected in previous seasons, e.g. new virulent races of cereal rusts. For soil-borne pathogens and slowly developing diseases of perennial crops, the relative abundance in the previous season is often a reasonable guide to the likely occurrence in the current season.

Some diseases attack specific parts of the plant such as immature leaves or growing berries. It is important to know what stages of the crop are susceptible as only these stages need to be protected.

Adequate forecasting can be achieved only if the biology and ecology of the pathogenic organisms is well understood. The dispersal and infection processes of many pathogens are dependent upon particular climatic conditions (e.g. rainfall, temperature) and most disease forecasting depends on predicting particular periods of weather likely to be favourable for infection. Examples of forecasting techniques are the attempts to predict outbreaks of apple scab (Mills periods) and potato blight (Beaumont periods) in Europe and North America which are favoured by periods of high humidity and temperature following rainfall.

Sigatoka disease of bananas is the only disease of tropical crops for which disease forecasting is attempted on a wide scale. The spores of the fungus causing Sigatoka leaf spot require rainfall and high humidity for their production, dispersal and infection. High temperatures also favour rapid development of the disease because the generation time between infection and the production of a new crop of spores is reduced and the epidemic progresses faster. An instrument called the Piche evaporimeter is used in the Caribbean to measure the relative dryness of the prevailing weather. The results are then coupled with temperature measurements and an assessment of the amount of leaf spot present on the banana plants at this time. The results are used to predict the safest intervals between applications of fungicides. Low incidence of leaf spot with high evaporimeter readings, such as occur in dry weather, indicate that the disease will be slow to progress and intervals between spraying can be large. When high incidence of leaf spot occurs with low readings, disease development will be rapid so that more frequent spraying is necessary.

The occurrence of most diseases is usually predicted on a seasonal basis, i.e. when weather conditions permit spread and infection by the pathogen, coupled with the known or predicted presence of the pathogen and the presence of susceptible host plants or stages of crop growth.

Further reading

Kranz, J., Schmutterer, H. and Koch, K. (1978). *Diseases, Pests and Weeds in Tropical Crops*. Wiley: London.

Part 2 Non-chemical methods for control of pests and diseases

5 Introduction

There are many different methods of controlling plant pests and diseases. Some are applicable to a wide range but most are effective only against a small range of pests and diseases which have similar methods of survival or dispersal between host plants. There are some individual pests or diseases that have a unique feature which has permitted the development of a special control technique.

In agriculture and horticulture, control of widespread pest infestation and disease epidemics in plant populations is of greater importance than control on individual plants. This means that low levels of pests and disease are acceptable, provided that no detectable loss of quality or quantity of produce occurs. A new pest or disease previously unrecorded in an area, or dangerous pests and diseases known to develop rapidly into serious and damaging epidemics, must be wiped out completely, if possible.

Factors influencing choice of method

Often a choice or combination of control measures is available and the decision of which to use depends on a number of factors relating to the circumstances under which the crop is grown. The identity of the pathogen and an understanding of its biology are the first essentials in this decision. Other considerations include the following.

1 The presence of other pests or diseases may favour the use of a method which would control several of these together.
2 Some methods may be more easily incorporated into the existing agricultural system than others.
3 Economic factors or the availability of special resources may restrict the choice of control method to one involving the least expenditure, specialised apparatus, chemicals or skill.
4 A high-value specialist crop or crop grown for seed may require a method which will give the maximum possible degree of control irrespective of other factors.
5 The predicted severity of the pest infestation or the disease epidemic may influence the choice. Where prevalent climatic or agricultural conditions indicate that a pest or disease may become severe, more effective measures will be needed than if losses due to pest or disease are expected to be small. Thus control methods, most of which need to be started before appreciable damage by pest or disease is apparent, may be regarded as an insurance premium and related to the risk involved.

Classification of non-chemical control methods

A commonly used system of classification is that which takes into account their mode of operation.

Exclusion A pest or disease is prevented from reaching a new area, or a new crop. Phytosanitary legislation, plant quarantine and the use of disease-free seed and vegetative planting material are examples of this.

Eradication A pest or disease is removed from an area or crop. For example, where an outbreak of a pest or disease occurs in a new area. Chemical or physical methods may be used to eradicate pathogens, e.g. sterilisation of soil; the use of fungicides on plant material or the heat treatment of seeds. There are various sanitary methods which destroy infested or diseased plants or crop debris and thus reduce the number of pests or the amount of pathogenic inoculum.

Avoidance Crops are grown in certain ways so as to avoid exposing them to pests or to disease infection, e.g. early sowing; sites free from pests and disease; the use of crops known to be immune.

Protection Protectant or eradicant chemical sprays are used on the plants. Plants with inherited resistance or tolerance to a pest or disease have built-in protection. Modification of the environment to protect plants against pests and disease is used in glasshouses and for the control of post-harvest pests and diseases. In addition agricultural crops may be grown under optimal conditions to avoid predisposing them to pest infestation and disease infection.

This classification is, however, based more on theoretical than on practical considerations and the methods have been grouped as follows:

1 international and national control measures, including the provision of clean seed and planting methods;
2 the breeding, selection and use of resistant plants;
3 cultural control methods aimed at reduction of pests and diseases in the field;
4 biological control methods;
5 chemical control methods, these are of such importance that they are discussed at length in Part 3;
6 integrated control methods;
7 pest management and/or eradication;
8 physical control measures, these are relatively unimportant.

6 International and national control measures

Pests and diseases have evolved to their fullest extent in the centres of origin of specific crop plants. Many pests and diseases are, therefore, localised in specific areas and it is in the interests of agriculture generally that their distribution should not be extended.

Fortunately, the range of crop species has been extended to new areas without spreading many of their pests and diseases. For example, *Hevea brasiliensis* (para rubber) evolved in the Amazon basin where it is attacked by the South American leaf blight, caused by a fungus, *Microcyclis (Dothidella) ulei*. When the crop was introduced into Asia this disease was not carried with it. Asia now produces almost 90 per cent of the world's rubber, largely because of the absence of this disease. Similarly, coffee which evolved in the forests of east and central Africa, thrived much better when grown in South America. This was because of the absence of coffee rust, *Hemileia vastatrix*, and important pests such as *Antestiopsis* spp. and *Hypothenemus* sp., the berry borer. Much of this advantage has now been lost, as both pests are now established in South America.

International phytosanitation and plant quarantine

Most countries have compiled lists of important pests and diseases which are absent from their crops. They have legislation forbidding or restricting the importation of particular crops or plants from infected areas. Generally these laws only permit the importation of insect-free and disease-free plants. Some plants, particularly those likely to carry dangerous diseases or pests, are completely prohibited. Other plants are permitted entry after rigorous inspection to ensure that they are pest-free.

Most countries are grouped in natural geographical units free from particular pests and diseases and with common geographical barriers to natural spread. It is in the interest of the countries in a unit to enforce common phytosanitary regulations.

Phytosanitation Most geographical zones have a regional organisation which coordinates the practice of phytosanitation in the member nations (Fig. 6.1 on p. 40).

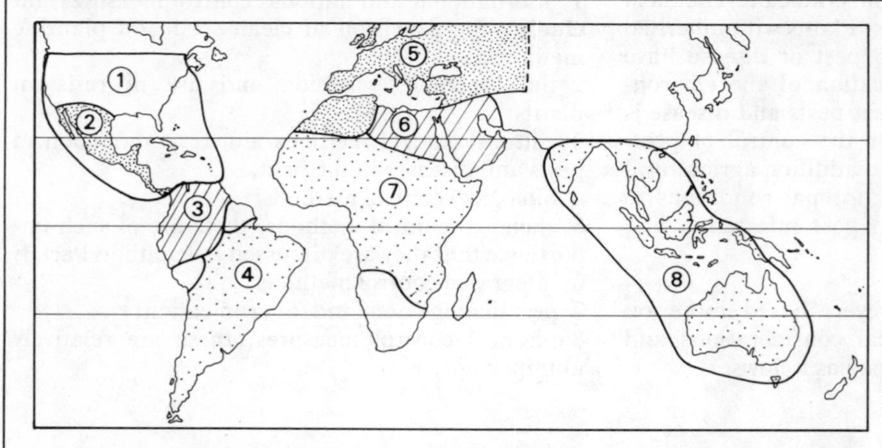

fig. 6.1 Map of phytosanitary areas throughout the world (after Mathys (1977))

The FAO established a Plant Protection Convention in 1953. This drew up sets of fundamental rules for a unified system of plant import and export. Many of these are incorporated into the International Phytosanitary Certificate (Fig. 6.2). This basic document is usually required for import of plant material. The exporting country is requested to certify freedom from certain pests and diseases and/or to specify that certain treatments, such as fumigation or seed-dressing have been carried out. Most countries, either themselves or in association with a regional plant protection organisation are signatories to this convention.

Most of the major crop diseases of the world can be carried on seed or vegetative planting material. For some diseases, e.g. viruses and smuts, this is the chief method of survival and of dispersal to new areas. This is of particular relevance to international phytosanitation. There is extensive international movement of the seed of new crop varieties. Many introductions of diseases to new areas can be traced to the importation of unclean seed material. Methods used to produce healthy seed and to detect diseased seed are considered later (p. 59).

Plant quarantine There are occasions when plant material that cannot be certified as healthy must be imported. For instance when new crop varieties are introduced. The improvement of collections of a particular crop (including cultivars and wild types) is sometimes necessary for breeding programmes, etc. Closed quarantine procedures can then be used. The plants are grown in biological isolation until freedom from pests and diseases can be ensured (Fig. 6.3). Quarantine stations should be remote from agricultural areas and need careful supervision.

fig. 6.3 Sugar cane plants growing in separate compartments of a quarantine house

MODEL PHYTOSANITARY CERTIFICATE

Plant Protection Service No.: ....

of: ....

To: Plant Protection Service(s)

of: ....

This is to certify that the plants or plant products decribed below were inspected in accordance with the requirements of the importing country and found free of quarantine pests and substantially free from other injurious pests; and that they are considered to conform with the current phytosanitary regulations of the importing country.

Identification and Description of Consignment

Name and address of exporters: ....

Name and address of consignee: ....

Number and description of packages: ....

Quantity and name of produce: ....

Distinguishing marks: ....

Place of origin: ....

Botanical name: ....

Means of conveyance (if known): ....

Port of entry (if known): ....

Disinfestation or Disinfection Treatment

Date: .... Kind of treatment: ....

Chemical (Active Ingredient): .... Duration and temperature: ....

Concentration: .... Additional information: ....

Additional Declaration:

Place of Issue: .... Date: ....

(Stamp of the Service)

....

(Name of Inspector)

....

(Signature)

No financial liability with respect to this certificate shall attach to (Name of Plant Protection Service) .... or to any of its officers or representatives.

fig. 6.2 International phytosanitary certificate

Third country quarantine occurs when the material is grown under controlled conditions in a country remote from either the exporting or importing countries. This has been used widely in distributing clonal material of tropical perennial crops throughout the world. The third country is usually a developed temperate country where pests and diseases of exotic crops pose no danger to indigenous crops. Specialised facilities, e.g. for the detection of virus diseases, will be available.

In some circumstances there will be a routine treatment of material on arrival in a country to eradicate any pests which may be present. Control for insect infestation often consists of fumigation of the plant material. Importation of deciduous fruit into many African countries from Asia and America is very rigorously controlled because of the danger of importation of San José scale (*Quadraspidiotus perniciosus*, Fig. 6.4). This is potentially the most destructive orchard pest. It caused tremendous damage to commercial orchards in the southern USA at the turn of the century. It attacked all pome fruit, plums, peach and cherries, in addition to shade trees and ornamentals. It was responsible for the death of many trees. Citrus fruit is checked for California red scale (*Aonidiella aurantii*) which can cause widespread damage (Fig. 6.5).

National control measures

In spite of international phytosanitary regulations, new pests and diseases do become established in countries previously free from them. These new introductions can be dealt with in several ways, all of which require the backing of legislation so that the measures can be enforced.

Eradication

This is usually the first line of defence taken against a newly introduced pest or pathogen. Unfortunately eradication is uncertain and difficult and may have important economic and social repercussions.

Detection of the foreign pest or pathogen at very low incidence levels must be infallible if eradication is to be complete. Dispersal to new areas may occur before the attacking species has increased sufficiently to be readily observable. Thus, detailed and frequent surveying is necessary.

Eradication should be carried out immediately after detection. It requires the destruction of the infected crop and surrounding areas of healthy crop into which the pathogen may have been dispersed. Crops are best destroyed by application of a desiccant herbicide and then burnt on site. Even the act of eradication may result in

fig. 6.4 Infestation of pears by San Jose scale (*Quadraspidiotus perniciosus*)

fig. 6.5 A young orange infested with California red scale

dissemination of the pest or pathogen to new areas. Compensation of farmers whose crops are destroyed and the provision of an alternative source of income is necessary.

There is no guarantee that eradication will be successful and continued vigilance is necessary to ensure that no escapes have occurred. The degree of technical resource required for successful eradication is frequently beyond the capacities of many developing countries, but coffee rust (*Hemileia vastatrix*) was successfully eradicated from Papua New Guinea in the late 1960s.

Eradication of alternate hosts of pathogens may be necessary, especially where these serve as an important overwintering source of the pathogen. Barberry is an alternate host to *Puccinia graminis*, the cause of black or stem rust of wheat; a barberry eradication campaign was set up in 1917 in the USA in order to help control this disease.

Prevention of spread

Newly established pests and diseases are often restricted to well-defined localities. Efforts to contain them can be made by preventing the movement of agricultural produce, etc. which may carry the pathogen out of the area (Fig. 6.6). The restriction of internal movement is frequently used in conjunction with attempted eradication.

The danger of the introduction of new pests is particularly common at ports where storage pests are the most common danger. *Trogoderma granarium* is potentially a serious pest in warehouses and godowns or food stores where it attacks a wide range of grains; all recorded species of *Trogoderma* are destructive of stored products (Fig. 6.7). Regular introduction of *Trogoderma* spp. occur in the harbour areas of Dar es Salaam and Mombasa and very strict control measures are taken to ensure that there is no spread of these infestations.

Other measures to prevent the spread of pests and pathogens include restrictions on the growing of susceptible crops in certain areas. This may act by creating a barrier or phytosanitary zone. It was used in an attempt to prevent the spread of coffee rust southwards in Brazil. It is still used to prevent the spread of the cotton red bollworms in East Africa. In the cotton-growing areas of Africa and the Middle East the red bollworms are serious pests which cause considerable damage to developing bolls; they are *Diparopsis castanea* — south of East Africa; *D. watersi* — north of East Africa. They are fortunately monophagous pests whose diet is restricted solely to cotton plants, and they have limited powers of dispersal. The main area adjacent to East Africa where red bollworm occurs is Malawi, but by the very strict enforcement of the law maintaining a cotton-free

fig. 6.6 Spraying vehicle wheels with fungicide to prevent carriage of coffee rust spores in Nicaragua

fig. 6.7 *Trogoderma* sp. larvae

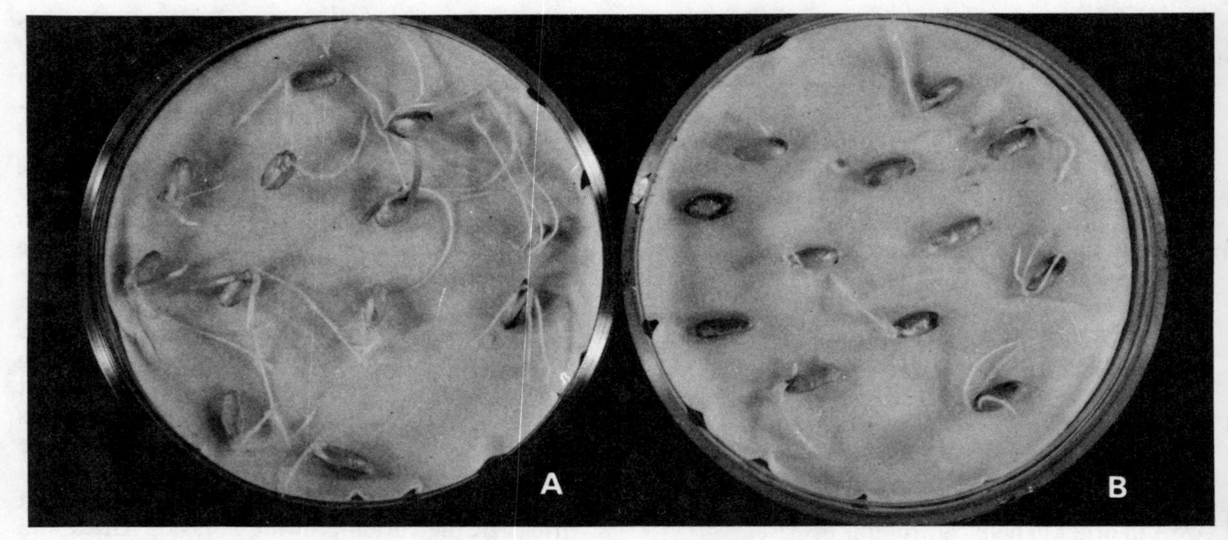

fig. 6.8 Simple test for seed health. Wheat seeds germinating on most filter paper: A is a healthy sample, B is a sample with diseased seeds (note dark areas indicating fungal growth)

zone in southern Tanzania this pest is prevented from extending its range into East Africa.

Restricted growing of susceptible crops may also act by starving out pests or pathogens from certain areas, or by providing a **close season** which is a break in an otherwise continuous food source for the pest or pathogen. Such measures are often used against specially injurious indigenous pests and pathogens.

Preventative methods may be used in combination but they all depend upon adequate ability to detect the pathogen. Technical and administrative resources are necessary to ensure that the various restrictions are strictly complied with.

Control of seed quality

The importance of healthy seed is discussed elsewhere (p. 59). The production and marketing of seed forms a distinct and specialised section of the agricultural industry in many countries. It is often under some form of national governmental control. This is to ensure the effective operation of quality control. This may take the form of a voluntary scheme whereby seed has to pass certain minimal requirements laid down by law before it can be marketed as 'approved' or 'certified'. The seeds would be tested by official methods for purity, health and germination capacity (Fig. 6.8). Government services usually provide the necessary expertise to inspect seeds but this may be done by an independent and impartial body. The International Seed Testing Association and the Danish Government Institute of Seed Pathology for Developing Countries have established recognised methods of testing the health of seeds; they also train personnel and advise on standards, etc. As a result the health of seed produced and sown in developing countries has been improved. The idea that only seed which was not fit to eat could be used for planting no longer prevails. The control of seed quality is an important national responsibility in maintaining or improving the health of crops.

Futher reading

Ebbels, D.L. and King, J.E. (eds). (1979). *Plant Health*. Blackwell: Oxford.

Hewitt, W.B. and Chiarappa, L. (eds). (1978). *Plant health and quarantine in international transfer of genetic resources*. CRC Press: Cleveland, Ohio.

7 The breeding, selection and use of resistance plants

All wild plants in natural ecosystems possess resistance to some pests and to most diseases. Man's domestication of some of these plants during the development of agriculture has resulted in their modification through selection and now special cultivated varieties or cultivars are recognised. An understanding of the mechanism of plant genetics has enabled considerable advances to be made in the production of new and better cultivars during the last 50 years or so. As mentioned in Part 1, advances in agricultural methods, such as monoculture and mechanisation, as well as changing economic circumstances, have increased the need for higher yields and quality and emphasised the importance of pests and diseases, many of which have become more severe under modern agricultural conditions. New cultivars have played a decisive role in the attainment of these needs and resistance to diseases in particular has always been an important selection criterion. The breeding or at least selection of suitable cultivars of major crops has been an important aspect of national agricultural development programmes. The genetic improvement of the world's major food crops is of such fundamental importance that much of this work is now carried out at the International Agricultural Research Centres which are funded by a number of international donors. The International Centre for Maize and Wheat Improvement, CIMMYT, based in Mexico, was responsible for the development of short-strawed, high-yielding tropical wheats and the International Rice Research Institute in the Philippines, IRRI, was responsible for developing the high-yielding IR rice varieties. The 'green revolution' of the 1960s resulted from the efforts of these centres.

The International Potato Centre in Peru, CIP, is responsible for potato improvement; IITA, in Nigeria and CIAT, in Colombia for improvement of root crops such as cassava and yams and legumes; and ICRISAT in India for sorghum, millet and dry area pulses. The maintenance of comprehensive germ plasm collections of the world's major crops is undertaken by the International Board for Plant Genetic Resources and this material is used by the international centres to breed and select many thousands of different varieties of crop species. The screening for pest and disease resistance is an important part of their programme.

Against insect pests

In large areas of agricultural crops there may be a few individual plants which either harbour far fewer pests or else show little sign of pest damage. These few individuals often represent a different genetical variety from the remainder of the crop. This variety is said to show resistance to the insect pest. When different varieties of the same crop are grown adjacently, either for different agricultural or culinary properties, differences in infestation level can be very marked.

Resistance to pest attack is characterised by a lower pest population density or fewer damage symptoms on the resistant plants.

Varietal resistance to insect pests is broadly classified in three categories: tolerance, antibiosis and non-preference. Some scientists restrict the use of the term varietal resistance to antibiosis, but non-preference can in practical terms be of even more importance in pest management programmes.

Tolerance The host plant suffers little actual damage in spite of supporting a sizeable pest population. This is characteristic of vigorous plants that heal quickly and show compensatory growth.

Tolerance is frequently a result of the greater vigour of a plant, and this may arise primarily because of the more suitable growing conditions rather than because of the genetic constitution of the plant. For example, sorghum growing vigorously will withstand a fair amount of stalk borer damage with no loss of yield. From a pest management point of view a tolerant variety can be a disadvantage in that it can support a larger population of the pest and so may encourage a build-up in local population rather than a decline.

Sometimes pest attack on a tolerant variety can actually increase the crop yield. This occurs quite frequently with the tillering of cereals following shoot fly, stem borer, or cutworm destruction of the initial shoot in the young seedling. So the use of a tolerant variety, in isolation, can be agriculturally advantageous.

Antibiosis The plant resists insect attack and has an adverse effect on the bionomics of the insect pest. The plants are characterised by thick cuticle, hairy stem and leaves, thickened stem, toxic sap, etc. Antibiosis has both biophysical and biochemical aspects. Morphological factors include hairyness of leaves and stems, tightness of husks in maize and tightness of leaf sheaths in rice, compactness of the panicle in sorghum, thickness of the stem in cereals, and also the diameter of the hollow pith in cereals.

Cotton jassids (*Empoasca* spp.) have ceased to be important pests of cotton in Africa since the development of pubescent strains which the bugs seem to find quite unacceptable as host plants. Similarly, hairy-leaved varieties of wheat in North America are significantly less attacked by the cereal leaf beetle (*Oulema melanopus*).

The tightness of the husk in some maize varieties has indirect effects which result in less damage to the cobs by corn ear worm (*Heliothis zea*) in the United States. At the International Rice Research Institute, IRRI, it was found that the tightness of the leaf cleavage in rice varieties was closely correlated with borer resistance. If the leaf sheath was tight and closed and covered the entire stem internode, then the young caterpillars frequently failed to establish themselves between the leaf sheath and the stem. They normally spend about 6 days feeding in this region before boring into the stem. Open-panicled varieties of sorghum in Africa are far less likely to suffer attack by false codling moth (*Cryptophlebia leucotreta*). The solid-stemmed wheat varieties are resistant to wheat stem sawfly (*Cephus cinctus*). Several species of pyralid and noctuid stem borers are restricted to stems of a particular thickness.

Anatomical factors are of less importance; the most pronounced is the development of silica deposits in the leaves and stems of graminaceous crops. Grasses are of a xerophytic habit and commonly have extensive silica deposits in the leaves and stems, and this development is not uncommon among cereal crops, especially those indigenous to semi-arid areas (e.g. sorghum). In sorghum plants silica deposits are apparently formed in some resistant varieties at about the 4th leaf stage. Up until this time all varieties of sorghum are attacked by sorghum shoot fly (*Atherigona soccata*), but infestations are not usually recorded in the resistant varieties once these silica deposits are formed, although infestations occur in the susceptible varieties up to the 6th leaf stage. Sorghum breeding programmes incorporate this type of resistance which is readily passed on genetically.

At IRRI, rice varieties have been developed with high silica contents in the stems and leaf sheaths. They are very resistant to moth larvae whose mandibles become worn down by the abrasive silica deposits.

Biochemical factors arise from differences in the chemical constituents of the plant. They may be restricted to particular parts of the plant body and/or particular stages of plant growth. It is thought that in some resistant plants the pest concerned suffers nutritional deficiencies resulting from absence of essential aminoacids. Some maize varieties show direct physiological inhibition of larvae of the European corn borer (*Ostrinia nubilalis*); they possess biochemical growth inhibitors at some developmental stages.

Non-preference Certain plants are less attractive to the pest for feeding or oviposition purposes because of their texture, colour, odour, or taste.

Non-preference is usually shown up when the ovipositing female insect refrains from egg-laying on the plant. In the case of chewing caterpillars or sap-sucking bugs it is shown that they do not feed when on the plant. In the Philippines, *Chilo suppressalis* females lay about 10–15 more egg masses on susceptible rice varieties than on resistant ones. At IRRI, the brown rice planthopper (*Nilaparvata lugens*) punctures the tissues of a certain rice variety but apparently does little feeding, possibly because of a reduced quantity of an aminoacid in the sap of that variety. In Uganda, experiments showed that cotton lygus bugs (*Taylorilygusvosseleri*) found some red-coloured cotton varieties less acceptable for feeding purposes than the usual green varieties; this difference in preference was attributed to the presence of certain aromatic compounds in the sap of the plants.

Varietal resistance has been shown in a broad range of crop plants including cereals, herbaceous plants, and trees, against an equally broad range of insect pests, including Hemiptera, Lepidoptera, Coleoptera, Diptera, and also Thysanoptera, in both temperate and tropical parts of the world. At present there are extensive breeding programmes in operation on rice (IRRI-Philippines), sorghum (East Africa), maize (India), pulses (East Africa and UK), cotton (Africa), etc. Most major tropical countries act as host for at least one crop plant breeding programme. The scope of these breeding programmes is often very large, for example at IRRI all 10 000 varieties of rice, representing the world germ plasm collection, have been screened for resistance to stalk borers. Twenty varieties show a high degree of resistance to *Chilo suppressalis* larvae, by antibiosis and/or non-preference, and some of these are also resistant to planthoppers.

In many respects plant breeding for insect resistance is one of the most ecologically desirable methods to be employed in pest management programmes. It takes some time but the results can be rewarding. It must be stressed that, in response to the development of resistant plants, it is probable that new insect biotypes will arise to which the existing plants are not resistant. However, most insect resistance in plants is complex and polygenic in nature which will discourage development of insect biotypes.

Against diseases

Parasitism is a specialised condition because plants are immune to most diseases, thus maize is not attacked by cocoa black pod, etc. Of the several thousand plant pathogens known, each plant species is attacked by only a few pathogens, and relatively few pathogens can parasitise a wide range of hosts. Furthermore, those pathogens which do attack a particular plant species do not, in natural plant communities, seem to exert a great influence on its survival.

In natural plant communities both devastating plant disease epidemics and complete absence of disease are rare. During the evolution of wild plant species, extreme susceptibility to disease has been eliminated by the process of natural selection. Similarly, pathogens unable to overcome absolute resistance (immunity) of the host have not survived. This very simplified view helps to explain why a balance between host and pathogen is achieved in natural plant communities, and why wild plants generally possess some resistance to most diseases.

Agricultural systems differ from natural plant communities in several ways which favour plant diseases, so that this balance is upset. For instance, monoculture favours the pathogen so that the natural level of resistance is no longer effective against increased disease pressure and the selection of plants on the criteria of yield and quantity may result in natural resistance being lost. However, the use of the intrinsic, heritable resistance present in wild plants which effectively controls diseases in nature is a useful way of controlling them in agriculture. For potentially it is a 'once and for all' method which would reduce continuing expenditure of time and money on recurrent control measures.

Development of resistant plants

There are several ways in which cultivars possessing intrinsic disease resistance can be produced. Selection of suitable plants from a variable wild population is probably the simplest and most direct way. With most crop plants it is necessary to combine disease resistance with many other characters such as high yield and good quality. This is usually carried out by an intensive plant breeding programme in which selected plants are crossed and suitable progeny selected. A new variety may be synthesised from a line of quite dissimilar ancestors.

The processes involved in such intensive plant breeding cannot be described in detail here, nevertheless, there are steps which must be undertaken.

1 Plants showing resistance to disease, or any other desirable character, must be detected before a logical plant breeding programme can be set up. The source may be in exotic varieties, or may be detected in existing populations by suitable techniques.

2 The selected parents must be capable of hybridisation with the production of viable progeny. Various techniques such as the removal of anthers or the use of male gametocides or male sterile cytoplasm may be used to induce hybridisation of plants which are normally self-pollinating (inbreeding).

3 The progeny must show suitable combinations of desirable characteristics. It is unfortunate that a high level of overall disease resistance is not easily coupled with high quality and yield in many plants.

4 There must be adequate techniques for detecting suitable levels of desirable characters in the progeny. This is specially important for the detection of disease resistance which may vary according to environmental conditions and must be related to the disease pressure to which the plant may be exposed in the field.

5 The combination of characters in the new variety must remain stable over successive generations, whether sexually or vegetatively reproduced.

fig. 7.1 Plants of wheat varieties showing variation in susceptibility to disease

The use of resistant varieties forms the basis of the control of plant disease on all of the world's major food crops such as cereals, legumes, root crops, etc. and many of the industrial or export crops such as cotton, sugar cane and cocoa. Major diseases of the past have been reduced to mere curiosities in many cases; sugar cane is an example of a crop where disease resistance has been effectively used in the permanent elimination of most problems. However, resistance to many of the cereal diseases is only temporary. Genetic variation in the pathogen results in the appearance of new races (or pathotypes) able to overcome particular sources of resistance.

A continuing effort is required by plant breeders to find new sources of resistance by producing new cultivars. In this way they may keep one step ahead of the pathogen. This is illustrated by the continuing work in breeding varieties of wheat resistant to stem, leaf and stripe rust (*Puccinia graminis, P. recondita* and *P. striiformis* (Fig. 7.1). Several hundred races of these pathogens have been identified based on their ability to overcome different genetic sources of resistance in wheat. Rice blast disease (*Pyricularia oryzae*) is also notorious for the great numbers of different pathotypes which occur in different geographic areas.

Disease resistance

There are many ways in which disease resistance can be classified as it varies considerably in the way it operates, the method of its inheritance (of particular importance to plant breeders) and how it behaves under field conditions.

Resistance mechanisms

Morphological Natural barriers to infection are present in every plant. The bark of woody stems and the cuticle of herbaceous parts may be thicker or tougher in resistant plants. The characteristics of the cuticle, particularly waxiness, have been shown to be correlated with resistance to disease in many plants. The silica content of the epidermis of the rice plant influences resistance to blast disease (*Pyricularia oryzae*). The tightness of bud scales, or of floral bracts, can prevent the entry of some pathogens and confer resistance to many smut diseases, for example. The internal morphology of the plant such as the presence of thick cell walls, can influence the rate at which a pathogen can grow through the tissues. Characters of the vascular tissue such as the ratio of xylem vessels to tracheids can influence the rate at which vascular diseases develop. Some morphological changes are brought about as a direct response to the presence of the pathogen. These include the production of tyloses to block off diseased xylem elements, and the thickening of cell walls by the deposition of suberin or lignin at sites where fungal hyphae attempt to penetrate.

Biochemical Some plants may possess a high degree of resistance to plant pathogens because of the presence of certain inhibitory substances in their cell vacuoles. Red onion varieties have a high level of anthocyanins in their cells which confer resistance to several bulb pathogens. Sometimes substances are produced by the stimulus of attempted invasion by the pathogen. These include gums and tannins. Trees often exude copious amounts of gum as a response to infection, e.g. phytophthora canker of citrus. Many sorghum varieties produce purple blotches consisting of cells full of tannins as a response to infection by a variety of leaf-infecting pathogens. Other biochemical defence mechanisms may be more subtle, and substances called 'phytoalexins' are produced in minute quantities which may kill the invading pathogen. Sometimes these result in the death of a few of the invaded cells so that a small fleck appears on the plant surface. This so-called hypersensitive reaction often operates against obligate parasites such as rusts. It seems that the successful pathogen must be able to by-pass at least some of these host defence reactions before it can cause a progressive disease.

The inheritance of resistance

Major gene resistance Resistance which is easily quantified and inherited in a simple fashion has been most widely used in breeding new cultivars. This type of resistance is usually controlled by a set of major genes which can be easily identified and manipulated in the breeding programme. These usually confer immunity or a high degree of resistance and have been widely used in producing resistant varieties of the world's staple crops, particularly against obligate parasites. The mechanism of this type of resistance is usually biochemical and often of the phytoalexin type. The disadvantage is that the ease with which it can be manipulated by the plant breeder is often mirrored by the ease in which pathogens can overcome it by the production of new races or 'pathotypes'. This is a result of the 'gene for gene' theory which states that for every major resistance gene in the host, there is a matching virulence gene in the pathogen which can overcome its effects. The ease with which pathogens can overcome the effect of major host resistance genes varies. Those which remain effective for longer are termed 'strong genes'. By incorporating many strong resistance genes into a host, disease resistance may be more durable. It appears that pathogens tend to become deficient in other respects, such as their ability to survive and spread, as they acquire more of the equivalent virulence genes to overcome the strong host resistance gene.

Polygenic resistance Other types of resistance are controlled by many different genes acting together and are inherited in a more complex fashion. This type of resistance is less easily overcome by changes in the pathogen but is difficult to breed into plants deliberately as the genes responsible and their various interactions are difficult to identify. Polygenic resistance may be either morphological or biochemical in action.

Resistance behaviour

Of greatest importance to the farmer and agriculturist is how the resistance behaves in the field. Will it be effective against all races of a pathogen and how long will it last?

Vertical resistance This is only operative against some races (pathotypes) of a pathogen and can be relatively easily overcome by changes in the pathogen population so that it is essentially temporary in nature. Typically vertical resistance is inherited in a major gene fashion so that the 'gene for gene' theory applies. It is the type of resistance that has usually been used by plant breeders in developing resistant cultivars.

The appearance of new virulent races of pathogens able to overcome vertical resistance has produced spectacular epidemics of cereal rust diseases, e.g. wheat stem rust in North America. It still causes much concern, even though careful monitoring of the pathogen population is undertaken. In East Africa, where stem rust is endemic and wheat is grown almost continuously, new pathotypes appear very quickly. Although vertical resistance is often temporary, it is nevertheless valuable and is most effective where the pathogen moves and reproduces slowly. In addition, where the life of the host crop is short, cultivars with different vertical resistance can be substituted quickly as required. The disadvantages of vertical resistance are most pronounced when used against rapidly developing diseases of perennial crops.

Horizontal resistance There is another type of resistance which is apparently universal in wild plants, and occurs to a lesser degree in some cultivars of crop plants. This is called horizontal resistance (generalised or field resistance) and is usually polygenic. It operates against all races of a pathogen and cannot be overcome by genetic changes in the host population, so that the gene for gene theory does not apply.

The main advantage of horizontal over vertical resistance is that it is permanent. Often horizontal resistance produces a degree of resistance difficult to detect in simple pathogenicity tests, but is always detectable by its effect on the development of disease epidemics. For this reason it has only been recognised and characterised by plant pathologists in the last 10-15 years. Horizontal resistance mechanisms tend to be quantitative in their effect and in the way they are inherited. They may be either morphological or biochemical in action, but are frequently influenced by environmental conditions and plant vigour.

Horizontal resistance is not amenable to manipulation by conventional plant breeding techniques. Research into the production of new cultivars of crops incorporating horizontal resistance is being carried out. Selection from relatively large out-breeding populations seems to be the best technique. It has a number of important advantages over vertical resistance, but is less easily used to produce new cultivars (Table 7.1).

Table 7.1 Comparison of vertical and horizontal resistance

Vertical resistance	Horizontal resistance
Operates against some but not all races of a pathogen.	Operates with equal effect against all races of a pathogen.
Easily breached by relatively simple genetic changes in the pathogen (the 'gene for gene' theory, p. 49).	Relatively stable although gradual changes may occur over a long period.
Confers immunity or a high degree of resistance on the host. This may be detected by simple pathogenicity tests.	Confers a lower degree of resistance on the host. this may only be detected by its effect on the development of plant disease epidemics
Simple inheritance involving major genes.	Complex inheritance which depends on the interaction of many minor genes.

The natural occurrence and agricultural value of the two types of resistance are related to their characteristics. As vertical resistance is temporary, it tends not to occur in nature in situations where diseases are endemic in continuously growing hosts, e.g. tropical evergreen perennials including the giant grasses such as sugar cane, or where there is little genetic diversity in the host population. It is also of limited use in protecting such crops against disease. Vertical resistance is more valuable where there is genetic mobility and heterogeneity of the host population, i.e. as in out-breeding plants or where the host population changes seasonally, e.g. with annuals, or in the case of some leaf diseases of deciduous perennials. This is termed the spatial and sequential discontinuity of susceptible host tissue. Horizontal resistance, however, is equally useful to all plants, but is most important in perennial plants or in-breeders where there is more continuity of susceptible host tissue.

Greater attention to how resistance affects the development of disease epidemics in plant populations rather than to the reaction of individuals in controlled tests, has led to an improved understanding of how losses are caused by disease. This has shown that the lower levels of resistance associated with horizontal resistance are quite tolerable for adequate control of many plant diseases as they substantially slow down the rate at which disease epidemics develop. Moreover, it is now known that the widespread cultivation of crop varieties with a common source of vertical resistance substantially increases the chances of selecting new virulent races of the pathogen; this is often called selection pressure.

Selection and use of resistant varieties

The replacement of many of the older traditional crop varieties, which often possessed a fairly high degree of horizontal resistance to indigenous diseases, by new high-yielding cultivars has often resulted in the upsurge of diseases previously of minor status on the older varieties. New techniques in plant breeding have also produced some unusual disease problems. The production of 'synthetic' hybrids (a controlled cross between selected male and female plants) which produce a very vigorous first generation, is often achieved by using plants in which only the female part is fertile for the female parent, so-called 'male sterile' lines. The southern corn leaf blight epidemic in the USA in 1971 was the result of the appearance of a new race of the pathogen *Drechslera (Helminthosporium) maydis* that produced a toxin to which the hybrid was very susceptible. The susceptibility to the toxin was inherited in the cytoplasm of the female parent — a factor common to all hybrid seed produced by use of a certain type of male sterile parent.

It is clear that there is great value in using resistant cultivars to control plant diseases even though this technique sometimes causes new problems. New ways of incorporating better disease resistance into new varieties are constantly being sought and although horizontal resistance is little used in breeding programmes at present, it does occur in some crops at levels high enough to control diseases (e.g. in sugar cane against most diseases and in maize against rust, *Puccinia polysora*). Techniques to minimise the disadvantages of vertical resistance are being tested. These include the use of 'multilines' in which a crop is a composite of different genetic lines each possessing different vertical resistance. This emphasises the discontinuity of susceptible hosts and aids the incorporation of 'strong' resistance gene complexes into cultivars.

The use of resistant cultivars should always be considered when the need for disease control arises. For many diseases, varieties possessing some resistance are known and are often listed in seed catalogues. For other diseases, existing cultivars may not have been tested, so that it is necessary for plant pathologists to conduct observation trials to assess their resistance under agricultural conditions. Geographical variation in pathogen virulence is widespread so that varieties listed as resistant in one country may not be so in another and vice versa. A check on reactions to indigenous pathogens is always advisable when new crop varieties are imported.

There are basically two ways of testing and selecting resistance to a disease.

1 By the use of artificial inoculation of pathogen inoculum, collected from the field or grown in

culture, into plants grown under standardised conditions in greenhouses, for instance. This is often a useful method for screening large numbers of plants rapidly. The disadvantages are that the inoculation technique and growing conditions may be so unlike field conditions that a false impression of disease resistance may be obtained and that some resistance mechanisms which are only operative in the field may be by-passed. It is also often difficult to detect horizontal resistance in such screening tests.

2 By observation of natural disease development on cultivars grown in replicated field trials. This method is most sensitive as infection is natural, and disease progress and yield can be measured. Seasonal fluctations in conditions suitable for infection and in natural inoculum levels often require repetition of these trials over several years in different sites; they therefore require more time, space and effort than simple greenhouse screening. Inoculum variation can be overcome by growing 'infector plants' among the trial plots. These are plants which have been artificially infected and provide a constant high inoculum to infect the trial. Interplot interference can cause difficulties in interpretation of results from variety trials where small plots are used, as partially resistant cultivars which may be free of disease when grown on a large scale, may become heavily infected from the large amount of inoculum produced by neighbouring highly susceptible cultivars.

Clearly there is no substitute for performance under full-scale field cultivation, so the determination of resistance is usually based on a combination of initial small-scale screening tests followed by large-scale field trials in areas where the disease is present over several seasons.

8 Cultural methods

The various cultural practices involved in crop production such as soil preparation, sowing characteristics, fertilisation, and cultivation of the growing crop, etc., can have a large influence on the incidence of some pests and diseases. They may increase the resistance of the plant to diseases, or hinder spread of pathogen through the crop.

Avoidance Disease control may also be achieved by selecting varieties which can avoid disease attack. These varieties have the ability to grow earlier, or later, or in a different ecological environment from others. Early-maturing cultivars often avoid the worst stage of epidemics of foliar pathogens which tend to occur late in the growing season.

Tolerance There are also cultivars which are said to be tolerant to a disease, i.e. they become infected but show reduced symptoms and little loss of yield. Cultivars tolerant to some virus diseases are commonly grown. However, they do little to reduce the incidence of the pathogen, and extensive use of tolerance could mask the presence of a serious disease which could threaten other crops.

Resistant varieties can only be useful if they are true to type, and there are many examples of apparent 'breakdown' of resistance in a cultivar which has subsequently been traced to unreliable seed. Besides simple errors in labelling or in mixing seed batches the genetic constitution of cultivars may slowly change over successive generations, and reference to breeders' foundation stock may be advisable. The advantages which crop varieties can confer not only on disease control but on general agricultural productivity can only be fully exploited if the varieties are provided by a reliable and efficient seed industry.

Further reading

Robinson, R.A. (1976). *Plant pathosystems*. Springer-Verlag: Berlin.

Russell, G.E. (1978). *Plant breeding for pest and disease resistance*. Butterworths: London.

Van der Plank, J.E. (1968). *Disease resistance in plants*. Academic Press: New York.

Phytosanitation

Removal of diseased plants

The most effective agricultural methods are the phytosanitary methods which destroy sources of disease inoculum or reduce the number of pests. The direct removal of diseased plants, or diseased

fig. 8.1 *Armillaria mellea* producing fruitbodies on an old tree. Old tree stumps are important sources of pathogens in plantations of perennial crops

parts of plants may control adequately many slowly spreading diseases. It is less effective against diseases which are easily dispersed and develop rapidly. In these cases the pathogen may have spread already to cause new infections by the time that the disease is noticed. However, attempted eradication is often the only practical method of controlling root diseases of forest crops and perennial crops, e.g. *Armillaria mellea* on tea, rubber, etc. (Fig. 8.1).

Diseased plants, including adjacent ones showing incipient symptoms, should be completely dug up and as much of the root system removed as possible. They should be destroyed by burning, preferably on site. The surrounding area may also be sprayed with fungicides as an added precaution if the disease is easily dispersed by spores.

Diseases affecting limited parts of woody plants such as cankers or die-back of branches can often be controlled by pruning without destruction of the complete plant. Diseased material should be destroyed and wounds treated with a fungicidal paint. Incidence of many fruit diseases can be reduced by continuously removing and destroying infected fruits as they appear. This prevents them from becoming sources of inoculum to infect other healthy fruits, e.g. cocoa black pod. Rubbish should be burned to remove hiding places and breeding sites (*Oryctes* spp.). Infested plants should be burned to kill insects inside them, e.g. coffee branches infested by *Dirphya* spp. Any resting stages of pests that may diapause or aestivate in the plants (larvae of pink bollworm in cotton) should be destroyed by burning where possible. The collection and burning of unpicked and fallen berries is regarded as the best method of controlling coffee berry borer (*Hypothenemus hampei*) in robusta coffee. There are in fact laws to enforce crop hygiene for the control of *Oryctes* spp. on the East African coast; this method of control is very effective when enforced.

Annuals Phytosanitary methods applied during the life of a crop are considerably less useful on annual than on perennial crops. There is insufficient time during the life of an annual crop for slowly spreading diseases to increase to really damaging levels. Extensive plant removal and pruning of annuals is rarely economical. However, some systemic diseases such as viruses and smuts can be controlled successfully by rogueing out infected plants as soon as they are seen and before appreciable spread of disease can occur. This is specially important in the production of seed crops where health may be of greater consideration than yield. In this case the value of the crop could make such control economical.

Phytosanitation practices applied to growing crops may result in some set back caused by extra damage. Although simple to apply they are labour intensive and may be expensive. It is necessary, therefore, to find out whether this effort is worthwhile to promote the future health of the crop.

fig. 8.2 Handpicking pests on young cotton plants in The Gambia

fig. 8.3 A grapefruit attacked by fruit flies; the other fruit on the tree are protected by raffia bags

Mechanical methods

Handpicking and killing of pests was probably one of the earliest methods of pest control (Fig. 8.2) and is still regarded as a profitable method for the removal of *Papilio* spp. caterpillars from young citrus trees. It is, however, probably unimportant in most countries because of the high cost of labour. *Dirphya* spp. larvae and *Apate* spp. adults boring in branches of coffee bushes can be killed by pushing a springy wire (e.g. a bicycle spoke) up the bored hole and spiking the insect.

Mechanical drags, which crush insects on the ground, have been used against armyworm (*Spodoptera littoralis*) larvae but this practice has long been abandoned. Banding on fruit trees is particularly effective against flightless moths, caterpillars and ants, which gain access to the tree by crawling up the trunk. Spray banding of coffee bush trunks is practised against adults of the white borer (*Anthores leuconotus*) and the moths of the tip borers (*Eucosma nereidopa*) which typically rest on the shaded lower parts of the trunk.

Tsetse flies (*Glossina* spp.) are caught and/or field-sampled in several parts of East Africa by hand, using cattle as bait and by the use of special bicycles with a decoy plate on the front. The earliest method of locust control was to herd the hoppers into a large pit or trench which was afterwards filled with soil.

On smallholdings in Malaysia it is common practice to place around large fruits of grapefruit, pumelo and jackfruit a woven bag made of grass or raffia which deters fruit flies (Tephritidae) from oviposition. Occasionally one fruit may be left unprotected (Fig. 8.3) so that the fruit flies will concentrate upon this fruit which can later be destroyed.

Secondary hosts

Many pathogens can survive on secondary (alternative) or alternate hosts. It is advisable to remove them from the close proximity of crops susceptible to diseases which they harbour. Common weeds often act as secondary hosts of pathogens of related crop plants, e.g. solanaceous weeds are hosts of *Pseudomonas solanacearum*, causing bacterial wilt of potatoes and tomatoes; graminaceous weeds are hosts of many diseases and pests of cereals, etc.

These hosts are important sources of many obligate pathogens such as viruses, mildews and rusts, and should be removed. Rusts often require alternate hosts to produce different spore forms before the life cycle can be completed. The barberry eradication scheme in North America (p.00) attempted to reduce the incidence of wheat stem rust by removing such an alternate host. However, as far as tropical rusts are concerned alternate hosts generally seem to play little part in disease epidemiology.

Indigenous trees often act as sources of root disease to perennial crops, and should be removed completely when new land is planted. Ring-barking causes root starvation and death of root pathogens. Standing trees should be ring-barked a few years before removal. All root systems should be grubbed out after the trees are felled and before crops are planted. Similar precautions should be taken when shade trees are removed from existing plantations, or when old plantations are to be replaced.

Alternative hosts are equally important to crop pests, and many pest species live on wild plants. They will only transfer to crop plants when the population reaches a high density. Most leafhoppers (Cicadellidae) and planthoppers (Delphacidae) that are pests, feed and breed on wild grasses in the vicinity of the paddy and cereal fields. Similarly, cotton lygus is often found in larger numbers on sorghum than on cotton.

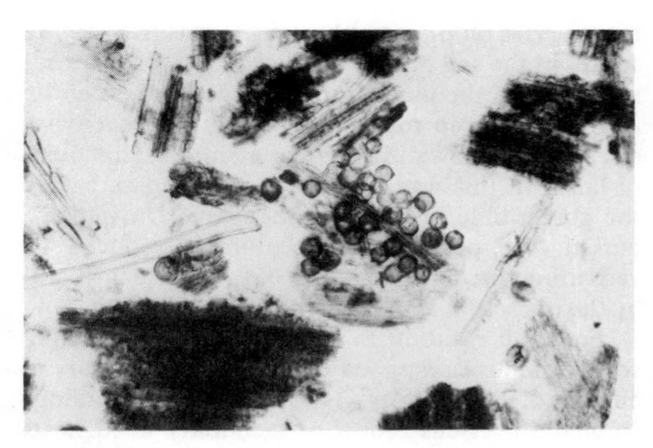

(a) Oospores of pearl millet downy mildew in leaf debris

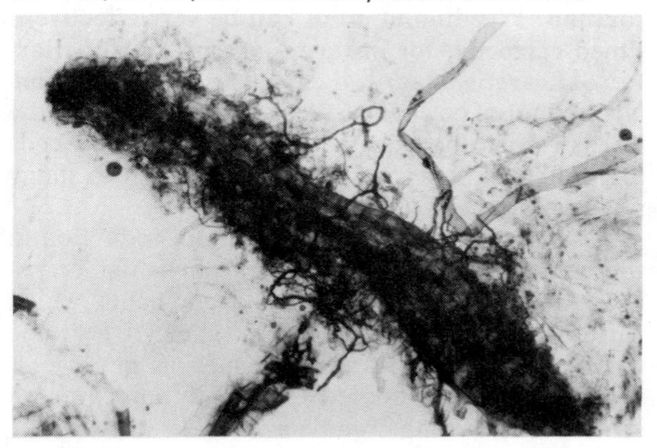

fig. 8.4 (b) Hyphae and spores of fungus in a piece of dead wheat root

Cultivation

The destruction of old plant residues between crops can be achieved easily during normal cultivation practices. It is a valuable phytosanitary practice.

Survival in plant residues is very often the main method by which diseases are carried over from one crop to the next (Fig. 8.4). So destruction of residue is of vital importance in controlling many diseases of annual crops. Bulky crop remains are best disposed of by burning. It may be possible to harvest them for an alternative use such as cattle feed or mulching, provided that this does not spread pests or diseases which may be in them. Deep ploughing may help to hasten the decomposition of crop residues.

Preparations of seed beds, harrowing for example, may result in the accumulation of small quantities of crop debris at the margins of fields. This can then be burnt or otherwise removed. Diseases such as bacterial blight of cotton, potato blight, etc. often start at the edge of fields where diseased crop residues have collected.

Crop residues can seldom be completely eliminated, but the inoculum available to infect new crops can usually be reduced to a minimum if suitable precautions are undertaken. However, even starting from a very low inoculum level, rapidly developing diseases can soon reach dangerous levels during favourable conditions.

Crop rotation

Diseases which survive on the remains of the host can sometimes be controlled by the destruction of the aerial parts of the plants between growing seasons. The soil-borne diseases, however, which may survive on pieces of host root or as free resting spores in the soil require further measures. Natural reduction of the inoculum of these diseases can be achieved by crop rotation in which different types of crops are grown in succession on the same piece of land. In this way several years may pass before the susceptible crop is grown again. During this period, the pathogenic inoculum is reduced by starvation due to the absence of a suitable host and by the microbial activity in the soil.

For crop rotation to be efficient in controlling soil-borne diseases, good cultivation is necessary to ensure that volunteers (stray self-sown seedlings from the previous crop), or secondary weed hosts do not allow the pathogen to survive during the rotation. Weeding of crops can be very effective if timed correctly, for instance, against chafers (larvae of Scarabaeidae) and cutworms (larvae of some Noctuidae). The eggs of these pests are laid on the weeds rather than on the crops, and if the weeds are pulled as the crop seedlings emerge then many potential pests are destroyed.

Crop rotation can also bring about reduction in the weed floras characteristic of particular crops. The different demands on soil nutrients which various crop have can also be balanced over the rotation period. Certain crops such as grass leys, legumes and deep-rooted crops may improve soil structure by adding organic matter. In this way the exhaustive effects of previous crops on the soil is counteracted.

Fallow periods during which no crops are planted may also be included in crop rotation and, if free from secondary weed hosts, may similarly reduce the incidence of soil-borne disease. Bare fallowing in dry seasons has been shown to assist in the control of soil-borne bacterial diseases and nematodes which are unable to survive exposure to hot dry conditions. Flood fallowing has been used to control panama disease of bananas caused by *Fusarium oxysporum* f. sp. *cubense*. In this process the land is flooded during a fallow period. The pathogenic inoculum in the soil is then destroyed by the particular microbial conditions, including a reduced oxygen supply, which the flooding induces. Shifting cultivation may be regarded as a form of long-term rotation since fertility increases and diseases decrease under the forest phase. The reverse happens when the land is cultivated. When these factors begin to limit production, the land returns to forest. Crop rotation and fallowing are also valuable methods for controlling many nematode diseases, particularly root knot nematode, *Meloidogyne* spp. which is a serious limiting factor to intensive vegetable growing on light soil.

Crop rotation separates crops in time but they may also be separated spatially. This can reduce the incidence of some plant diseases. Crops which can act as alternative hosts of the same disease should be kept apart. For example, maize can be severely damaged by sugar cane mosaic virus; cucumber mosaic virus can damage peppers (*Capsicum* spp.) and many other crops besides cucurbits. Under extensive monocropping systems one crop may extend continuously for hundreds of square kilometres. This allows diseases to spread rapidly with no natural check to dispersal. In smallholdings or peasant agriculture, however, dispersal is restricted by growing different crops in adjacent small plots. The rapid build-up of extensive disease epidemics is thus hindered.

Against monophagous or oligophagous pests crop rotation can be quite effective, especially in the case of insects which take several years to develop. It is not effective, however, against migratory pests or those with very efficient powers of dispersal. It is probably effective against chafer larvae (Scarabaeidae) and cutworms (Noctuidae). The alternation of cereals with non-cereals may be an important method of controlling *Nematocerus* beetles (Curculionidae). In most parts of Africa crop rotation is not an important method of control of insect pests; it is far more effective against nematodes and plant diseases under local conditions.

Disease avoidance

Diseases can be avoided by deliberately growing crops in areas known to be disease free. Shifting cultivation has this effect. Soil-borne diseases build up during the cropping cycle, but these diseases often do not occur when a new area is planted up. Crops may also be grown where climatic conditions are unfavourable for the pathogen, e.g. high altitude maize in Africa is not attacked to any extent by maize rust, *Puccinia polysora*; coffee at high altitude is less prone to rust, and in low areas less prone to coffee berry disease. In seasonally arid areas, irrigation may allow out of season cropping so that seasonal epidemics may be avoided. Thus various leaf blights of many vegetable crops can be severe during the cool moist winter months of North Africa and the Middle East, but crops grown under irrigation during the hot dry summer remain disease free.

Optimal cultural conditions

There are many agricultural practices which may increase the resistance of crop plants to disease or improve their tolerance so that losses are reduced. Very many diseases are more severe, or at least cause more damage, on plants which are growing under unfavourable conditions. Optimal conditions promote plant vigour so the physiological defence mechanisms may be more efficient. Rapid growth of young plants may enable them to grow away and recover from seedling diseases. Waterlogging, drought, shortage or imbalance of nutrients, unfavourable temperature regimes, etc. can predispose plants to disease. Frequently the effect works through a chain of interactions the net result of which favours disease development. Many examples of the varied effects of agronomic conditions on pests and diseases are known.

Time of sowing By sowing early, or sometimes late, it may be possible to avoid the egg-laying period of a pest. Timing can be managed so that the vulnerable stage in plant growth has passed by the time the insect numbers have reached pest proportions. Early sowing is regularly practised in many areas as a control measure against cotton lygus (*Taylorilygus vosseleri*) and sorghum midge (*Contarinia sorghicola*). Simultaneous sowings of the same crop over a wider area can be used to avoid successive plantings which often permit the building up of a very large pest population and favour rapid development of plant disease epidemics. Generally, the earlier a crop is sown the earlier it matures and the more likely it is to avoid serious damage by epidemic plant diseases which tend to reach a peak late in the season.

Time of harvesting Prompt harvesting of maize and beans may prevent these crops from becoming infested by maize weevil (*Sitophilus zeamais*) and bean bruchid (*Acanthoscelides obtectus*), respectively. Both of these pests infest the field crops from neighbouring stores but are not generally able to fly more than about 0·5 km. It is thus recommended that these crops are grown at least 1 km away from the nearest grain store. Harvesting during dry conditions helps to avoid grain damage by moulds and fruit rots. Early-maturing varieties may enable crops to be harvested before disease epidemics reach dangerous levels.

Close season In East Africa there is legislation to ensure that there is a close season for cotton growing in order to prevent population build-up of pink bollworm (*Pectinophora gossypiella*) which is almost monophagous on the cotton plant. The legislation stipulates that all cotton plants should be uprooted and destroyed (or burned) by a certain date; then no seed would be planted until the following rains arrive. However, it is obvious that many peasant farmers do not bother to destroy the old plants by the appointed date and so in some areas there is considerable survival of diapausing pink bollworm larvae. Similarly there is a natural close season for beans against the bean fly (*Ophiomyia phaseoli*) and it is strongly recommended that this crop should not be grown during the dry season, for at this time pest damage to the beans is at its greatest.

Special cultivation methods Deep ploughing may bring lepidopterous larvae and pupae, and beetle larvae, up to the surface to be killed by predators or inclement weather. A period of fallow usually will reduce the number of most soil pests.

Trap crops The use of trap plants to reduce pest infestation of various crops is based upon the knowledge that many pests prefer to feed on plants other than those on which they are the most serious pests. Trap plants are either grown as a peripheral band or interplanted at about every 5th to 10th row of the crop plant. This preference may be exploited:

1 the pests are lured from the crop on to the trap plants where they stay and feed;
2 only the trap plants are sprayed with pesticides because greater numbers of the pest are feeding on them than on the crop plants. Since there are many fewer trap plants the cost of spraying is reduced.

Soil-borne pathogens may sometimes be controlled by trap crops which stimulate the germination of resting spores or the similar structures by which they survive in the soil. The crop is then ploughed in or otherwise destroyed at an early stage before the pathogen has reproduced or formed new resting spores. Nematode control has also been attempted in this way but in general this method is seldom used. **Green manure crops** may act in a similar fashion against many soil-borne fungal diseases. These crops are grown for a short period only, usually until they have produced a reasonable vegetative growth. They are then ploughed in to the soil, considerably increasing its organic content. As a result microbial activity is increased and the destruction of pathogenic fungal inoculum may be hastened. Various parasites of nematodes, including specialised nematode-trapping fungi are also increased and may help to reduce the incidence of nematode diseases.

At Namulonge Cotton Research Station, Uganda, it has been shown that *Cissus* (Ampelidaceae) is a very attractive trap plant for lygus bugs on cotton. As little as 1 row in 10 of *Cissus* produces significantly lower lygus infestations on the cotton plants. Work has been done at Makerere University to determine the effectiveness of maize and other plants as traps for late bollworm of cotton (false codling moth) in Uganda.

Drainage Root and crown rots and damping off of seedlings are often increased by waterlogging, e.g. phytophthora diseases (Fig. 8.5), or by high soil temperatures, e.g. fusarium crown rots of cereals.

Fertilisers Excessive nitrogen may result in very luxuriant growth which can predispose plants to mildews and other foliage diseases. Adequate levels of phosphorus and potassium are often associated with greater resistance to root and vascular pathogens. High levels of sodium or soluble aluminium in the soil may be toxic to the plant and lead to leaf scorching and predisposition to leaf disease. Suitable treatment with fertiliser normally results in vigorous growth of the plants and this often causes them to be tolerant of pest damage.

Good husbandry High humidity within crops is a factor often contributing to severe outbreaks of foliage diseases. Adequate spacing between plants, or adequate pruning of perennial crops, and fertiliser regimes which prevent excessive luxuriance of foliage can reduce this effect. Foliage or plant density may also influence the prevalence of pests and predators and the incidence of virus vectors.

fig. 8.5 Death of transplanted seedlings caused by root rot influenced by poor cultural conditions

Clean seed material

Seed or vegetative propagating material such as tubers, cuttings, etc. is of basic importance in the production of healthy crops. It is the quality of seed material which determines crop potential; the realisation of this potential depends largely on cultural practices such as fertilisation, tillage, etc. Field husbandry techniques cannot improve bad seed material. Unfortunately, good seed can be wasted by poor husbandry; this has often led to loss of investment. Seed material should be genetically suitable, i.e. the correct cultivar for prevalent conditions should be selected, it should breed true to type, and have a reasonably high germination rate. The seeds must also be free from pathogens and pests.

Healthy seeds can only be produced by a healthy parent crop. The tolerable limits of disease in crops intended for seed production are very small; intensive pest and disease control measures are therefore required. Inoculum from neighbouring diseased crops may produce latent seed-borne infections on seed crops, e.g. smut and virus diseases. The health of adjacent crops must be taken into consideration. Contamination of the seed may also arise during harvesting, and seed health may be considerably impaired under bad storage. Seed with poor germination characteristics, even though it may be free from specific diseases, gives rise to a thin, weak crop in which soil-borne diseases can be particularly damaging.

Assessment of the health of seeds can be made by direct observation of seed material. Some knowledge of its origin can reveal much useful information. A simple germination test of samples on moist absorbent paper under uniform conditions can reveal the presence of many pathogens. Samples should always be examined by a competent plant pathologist, as the assessment of specific seed-borne diseases may demand the use of special techniques. This applies particularly to vegetative planting material where the detection of virus in, for example, potatoes, citrus root stocks, etc. requires screening with indicator plants and other materials. There are many techniques for assessing the health of seed, details of which should be sought in the relevant specialist literature.

Common soil pathogens frequently infect seed and planting material. It is technically impossible to achieve complete freedom from these. Nevertheless, the seed can be treated with heat or chemicals to eradicate many pathogens. Chemicals can also be used to protect the seed and young seedling plants against most soil-borne diseases and a few air-dispersed pathogens, during the early phase of the crop cycle.

Special methods are often used to produce disease-free foundation stocks which could not be used on a large scale directly for commercial seed. These methods are used principally against virus diseases of vegetatively propagated crops and include a variety of heat treatment methods, grafting on to special virus indicator plants so that healthy plants can be selected, and meristem tip culture. This technique involves excising the apical meristem of the plant (which is usually virus-free in mildly infected plants) and growing it on a special nutrient medium to produce a new healthy plant, from which subsequent propagations can be made.

Further reading

Garrett, S.P. (1970). *Pathogenic root-infecting fungi*. CUP: Cambridge.
Neergaard, P. (1977). *Seed pathology*. Vols I and II. Macmillan: London.

9 Biological control

Under natural conditions most insect populations are controlled by a complex of predators, parasites and pathogens that share the same habitat and belong to the same ecological community. This is sometimes called the **balance of nature**, or alternatively **natural control**. Agricultural systems tend to upset the ecosystem and create conditions suitable for a population explosion of pests and diseases of the crops grown. Sometimes the natural enemies of the crop pests can also increase in population density. They can then still exert a controlling effect on the pest populations. This is particularly true for orchard and plantation crops which are present for a long enough time to allow stabilisation of insect populations. With field crops such as cotton and cereals there is generally insufficient time available for the predator/parasite populations to build up to a level at which they exert any appreciable suppressive effect on the pest population.

Indiscriminate use of chemical insecticides has often resulted in the death of many natural enemies in excess of the pests the insecticides were aimed at. The pests sometimes increase in numbers despite the successive insecticidal sprays because the natural control has been removed.

It is sometimes necessary to supplement the natural control by deliberate introduction of predators, parasites, or pathogens into the area concerned — this is termed **biological control**, in the strict sense. A great deal of work has been done on this subject by the various stations of the Commonwealth Institute of Biological Control (CIBC), which are situated in Africa, India, Switzerland, Austria, Pakistan, Mexico and Trinidad.

Biological control can sometimes result in permanent control from one application, although it is usually slow in action. Generally this method is most effective on exotic crops which often do not have their full complement of pests in the locality to which they have been introduced. In addition the local predators, parasites and pathogens will invariably be in a state of delicate balance in their own environment and cannot be expected to exercise much control over the introduced pests. It is best to introduce the controlling agent from the native locality of the introduced plant.

Many biological control projects in the Tropics have been spectacularly successful, and usually at a cost far lower than would be required for an insecticide programme. Relatively few pests can be controlled biologically. It must be stressed that in all pest control programmes the first objective should be to refrain from harming the existing natural control balance.

The general cost of biological control programmes is difficult to estimate in many cases, but, for example, it has been reported that the Department of Biological Control of the University of California spent over the period 1923-1959 a total of \$3.6 million which resulted in a saving in California of about \$100 million.

Predators

Birds are of great importance as insect predators, although few studies of their effectiveness have been made. However, chickens are used in small cotton fields in parts of Africa and they give good control of cotton stainers. A recent description of biological control of rice pests in south China included the herding of 220 000 ducklings through the rice fields. The ducklings eat about 200 insects per hour, and most of these are pests; the use of these ducklings reduced the use of chemical insecticides on the early rice crop from 77 000 kg in 1973 to 6 700 kg in 1975. Fig. 9.1 shows pied crows eating armyworms in Nairobi.

fig. 9.1 Pied crows are predators of insects which attack crops; here the crows are eating armyworms

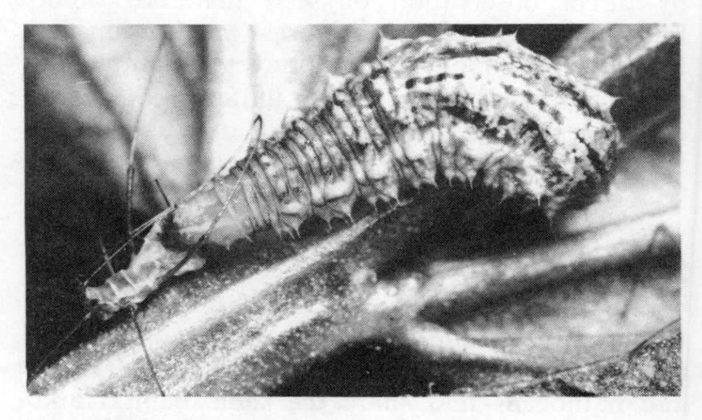

fig. 9.2 Larva of syrphid attacking the pea aphid *Macrosiphum pisi*.

Invertebrate predators include spiders, but their effects have not been studied to any great extent. Probably the most important insect predators are various Coccinellidae, both as larvae and adults, and the larvae of many species of Syrphidae (Fig. 9.2). These collectively prey mostly on aphids and scale insects, but also take psyllid nymphs, some other bugs and sometimes small caterpillars. Phytophagous mites are generally preyed upon by predacious mites, and in some countries a measure of control has been achieved using *Phytosielus* spp. mites as predators on *Tetranychus* spp. Fig. 9.3 shows ladybirds eating aphids.

Parasites

The most important group of insect parasites are the Hymenoptera (Ichneumonidae, Braconidae, and Chalcidoidea) which attack the eggs, larvae, pupae and sometimes adults of many other groups of insects (Fig. 9.4). Certain Chalcidoidea (especially Trichogrammatidae) are solely egg parasites. Some Diptera are important parasites of grasshoppers, and the Tachinidae are larval parasites of Hemiptera and grasshoppers. Important tachinid parasites in Africa are *Exorista sorbilans* on *Ascotis* sp. (coffee giant looper), and *Bogosia rubens* on *Antestiopsis* spp. (coffee antestia bugs); both species are quite common.

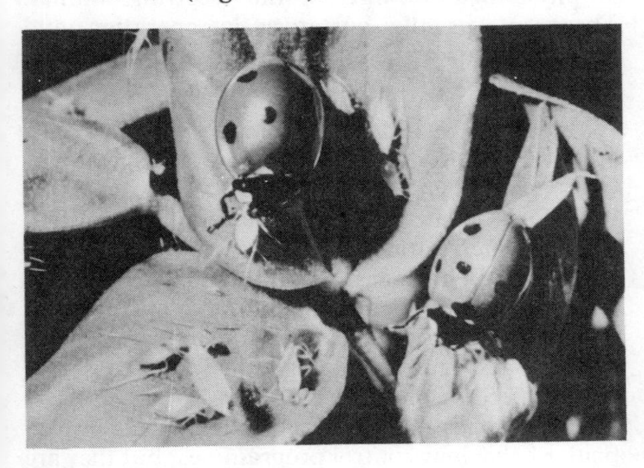

fig. 9.3 Ladybirds as predators of aphids

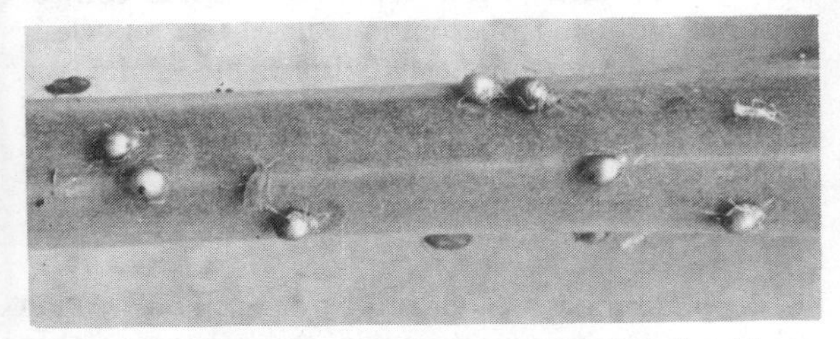

fig. 9.4 Aphids (*Myzus persicae*) parasitised by chalcid Hymenoptera on stem of Chinese cabbage

Pathogens

Fungi are not of general importance insecticidally at present. About 300 antibiotics show promise as pesticides, these act directly as killing agents or inhibitors of growth or reproduction.

Bacteria are more promising and various strains of *Bacillus thuringiensis* are effective against many caterpillars, and *B. popilliae* against some beetle larvae. Some proprietary formulations are now commercially available. It has been reported that a bacterial agent (*B. thuringiensis*) was responsible for a high mortality of the coffee berry moth *Prophantis smaragdina* in Tanzania.

Viruses are the most common pathogens attacking insects and have been most effectively employed against Lepidoptera (e.g. armyworm in East Africa, (Fig. 9.5), some Hymenoptera (sawflies) and a few beetles, *Melolontha* spp., and *Oryctes* spp.). There is some evidence from other countries that viruses may be effective against some red mites (Tetranychidae).

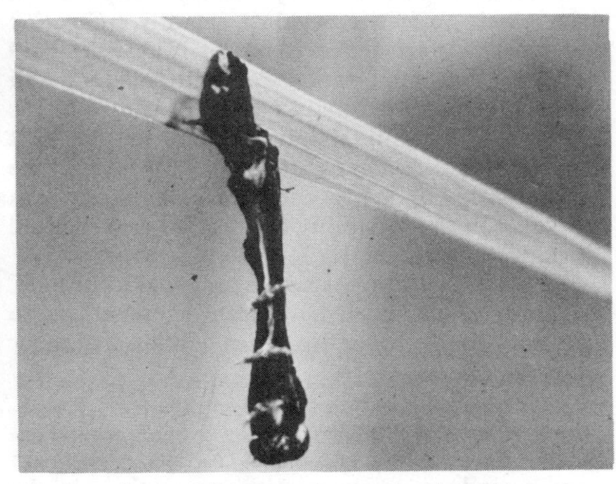

fig. 9.5 Larva of *Spodoptera exempta* attacked by a virus

Protozoa and nematodes have not been studied extensively for their potential in biological control of crop pests, but nematode control of mosquito larvae is well established. Control by use of pathogens is sometimes referred to as **microbial control**.

Sterilisation

This usually refers to the sterilisation of males by X-rays or γ-rays and is called the **sterile male technique**. Control of a pest by this technique is termed **autocide**. Sterilisation can also be effected by exposure to various chemicals and this practice is called **chemosterilisation**.

Male-sterilisation is effective in species where females only mate once and are unable to distinguish or discriminate against sterilised males. For example, on the island of Curacao in 1940 male flies of screw-worm (*Callitroga* spp.) were sterilised by exposure to γ-rays and dropped from planes at a rate of 154 per square kilometre per week. The entire pest population was eradicated in 12 months. The lifecycle took about 4 weeks to complete, and each female mated only once.

Generally, autocide is only really effective when applied to restricted populations on islands, etc., but can be effective on parts of continents. The screw-worm eradication campaign was extended to the southern part of the USA where the pest is very harmful to cattle, and in Texas 99·9 per cent control was achieved in only 3 years. Male sterilisation trials were effective against medfly (*Ceratitis capitata*) on part of the island of Hawaii in 1959 and 1960, but immigration from untreated parts of the island prevented maintenance of control. This method of control should be effective against *Oryctes* spp. attacking the coconuts along the coastal strips of Kenya and Tanzania. Work is in progress on the feasibility of autocide as a method of controlling tsetse.

Chemosterilisation is now a practical technique. A variety of chemicals have been demonstrated to interrupt the reproductive cycles of a large number of insect species.

Genetic manipulation

The electromagnetic radiations (X-rays, γ-rays) are used to induce dominant lethal mutations in the germ cells of the insects. These mutations in insect sperm have been used successfully in several eradication programmes. Lethal mutations are not lethal to the treated cell itself only to its offspring which fail to develop to maturity. These mutations arise as a result of chromosome breakages in the treated cells.

Pheromones

Pheromones (also called ecto-hormones) are those chemicals secreted by an animal into the external environment to elicit a specific reaction in a receiving individual of the same species. Pheromones have been divided into 2 groups according to their mode of action:

1 those which give a releaser effect entailing a more or less immediate and reversible effect on the behaviour of the recipient;
2 those which have a primer effect starting a chain of physiological events in the receiving animal. These are usually gustatory in operation and typically control the social behaviour of such insects as the Hymenoptera and Isoptera.

Behaviour-releasing pheromones are typically odorous and they act directly upon the central nervous system of the recipient. The types of behaviour elicited include alarm, aggregation, dispersal, territoriality, sexual activity, and trail following. Behavioural control with several different types of pheromone may be possible but recently attention has been focused upon pheromones causing aggregation for purposes of mating, feeding or oviposition.

Pheromones may be used by themselves in particular behaviour control programmes, but they are also of use in the carrying out of population density surveys. To date, however, no successful control programme using pheromones has been reported, although this technique is being explored for the possible control of locusts and mosquitoes, and of red bollworm (*Diparopsis castanea*) in Zimbabwe.

The two types of pheromones causing aggregation are:

1 sex pheromones produced by one sex only which trigger behaviour patterns in the other sex that facilitate mating;

2 general aggregation pheromones which may be produced by one sex only but cause aggregation behaviour by both sexes of the species.

Sex pheromones These are often referred to as 'sex attractants', or 'sex lures'; but the response to a sex pheromone is often a very complicated sequence of behaviour. The male reaction to the female odour is far more complex than a simple attraction to the odour source. Usually the sex pheromone is released by the female insect to guide the male into her vicinity. Thus behavioural control programmes using sex pheromones are difficult in that the male insect is attracted and not the egg-laying female. Certain male insects do release sex pheromones but these are usually only short-range mating stimulants. An exception to this rule is the cotton boll weevil (*Anthonomus grandis*) in which the males release a pheromone which attracts the females from some distance.

Aggregation pheromones It has recently been demonstrated that bark beetles (Scolytidae) invading the phloem tissue of suitable host trees produce pheromones which causes an influx of males and females of the same species into the same area. In some species (*Dendroctonus* spp.) the pheromones are produced by the female and in some *Ips* spp. apparently by the males.

Direct biological control of plant diseases has been little used so far because direct manipulation of the microecology of plant surfaces is very difficult. It is known that the micro-organisms which live saprophytically on leaf surfaces (the phylloplane) and close to roots (the rhizosphere) have a profound effect on the infection process of pathogenic organisms, and control of some diseases has been achieved on an experimental scale by adding certain antagonistic organisms to root or leaf surfaces. The injudicious use of fungicides, particularly if applications are badly timed, may increase the incidence of some diseases because of a reduction in antagonistic micro-organisms; this happened with some early attempts to control coffee berry disease.

The incidence of some soil-borne diseases including nematodes, may be reduced by using certain soil amendments or cultural practices which influence the microbiological activity of the soil and the rhizosphere. Green manure crops and organic compost both enrich the soil microflora which as a result becomes more antagonistic to such common soil pathogens as *Rhizoctonia* spp. and *Macrophomina* spp. Prolonged cereal cultivation without a break crop over several years results in a decline of some root diseases such as take-all, caused by *Gaeumannomyces graminis*. These processes may be regarded as indirect forms of biological control.

Examples of more direct control are the inoculation of tree stumps with *Peniophora gigantea* which is antagonistic to and prevents infection by *Fomes annosus*, and the cross-protection of high-value horticultural crops against virulent virus strains by inoculation with avirulent ones. Thus tomatoes are inoculated with avirulent strains of tobacco mosaic virus to protect them from virulent strains of this disease.

Further reading

Baker, K.F. and Cook, R.J. (1974). *Biological control of plant pathogens*. W.H. Freeman and Co.: San Francisco.

De Bach, P. (1974) *Biological control by natural enemies*. CUP: Cambridge.

Huffaker, C.B. (ed.). (1971). *Biological control.* Plenum Press: New York.

Kilgore, W.W. and Doutt, R.L. (eds). (1967). *Pest control – biological physical and selected chemical methods*. Academic Press: New York.

10 Integrated control

Early pest control measures at the beginning of the century were mainly concerned with the biology and ecology of pests, in particular as they relate to the numbers of pests. Attempts were made to make the environment less favourable to the pests by cultural and biological means. The latter included importation of insect predators and parasites in the earliest cases of biological control. The chemical poisons available at that time were simple, usually inorganic salts and kerosene, and neither very effective nor particularly persistent. Then in 1940 DDT became available as an insecticide, small doses of which are capable of killing a wide spectrum of insects, and with a long-lasting residual activity. Control of insect pests was revolutionised, and a series of new synthetic organic insecticides was rapidly discovered. However, control of pests continued to be a problem and often became a bigger problem than before. Many undesirable side effects were noted from the chemicals, in particular the accidental destruction of natural enemies of the pests, and the development of resistance to the insecticides by many pests. The cycle of events is now completed as we are turning back to the original concern with biological and ecological understanding. The judicious use of the synthetic organic insecticides is an important weapon in pest control, but must be used in conjunction with other appropriate measures.

The term **integrated control** originally described the combination of biological control with compatible chemical application. In this sense the term **biological control** included both natural control and biological control (in the strict sense). Chemical insecticides were to be used judiciously to avoid disruption of the natural control by killing the predators and parasites in the crop community. This can be done in several ways, by using specific, carefully screened insecticides only, by careful timing of treatment, by using minimal dosages, by reducing spray drift, and so on. This attitude was logical because these two approaches are our primary resources in the fight to control insect pests but in many instances they have been in direct conflict with each other.

However, practical experience and logic have made it clear that we must integrate not only chemical and biological control but all available procedures and techniques. In this way a single pattern aimed at profitable crop production will cause minimal environmental disturbance. This realisation led to the concept of **pest management** with its emphasis on the broad ecological approach to control. The literature on pest control is very confusing for in some cases integrated control is regarded in the old context and in others it means the same as pest management. To avoid this confusion some authors are using the term **integrated pest management** and are regarding integrated control as a historical term which has now been superseded.

However, the old idea of integrated control is still applied in some cases. The control of coffee leaf miners (*Leucoptera* spp.) at Ruiru in Kenya is a carefully balanced integrated programme using a few selected insecticides together with natural control by parasites may continue without interference. borer (*Eucosma nereidopa*). Here the shaded tree trunks are sprayed to kill the resting adults; this avoids the application of chemicals to the leaves and shoots so that the quite effective natural control by parasites may continue without interference. It was found in Sabah that excessive use of broad-spectrum pesticides severely aggravated the local pest problems on cocoa and oil palm plantations. The most effective immediate measure was to curtail spray applications in order to allow survival of the predators and parasites which normally kept most of the pest populations in check.

The importance in an area of the natural control exerted by the predators and parasites on the pest insects cannot be over-emphasised. Unfortunately the extent of natural control is often not appreciated until careless insecticide applications kill the predators and/or parasites and this is then followed by a pest population explosion. Typically leaf miners and scale insects (Coccoidea) are better controlled by their parasites than by chemicals, and most control programmes on coffee and citrus take this into account.

11 Pest management and eradication

Pest management

Pest management as a concept is now well established but the best definition is difficult to determine. It has been stated that integrated control originally referred to the modification of chemical control in order to protect and enhance natural control. The concept of integrated control has been extended and now can be used to embody most of the essentials of pest management.

Pest management can be defined as the reduction of pest problems by methods based on a thorough understanding of the life systems of the pests. It involves accurate prediction of the ecological and economical consequences of these methods in the best interests of mankind. In developing a pest management programme, priority is given to understanding the role of intrinsic and extrinsic factors in causing seasonal and annual changes in pest populations. It is necessary to understand the life system of the pest in relation to the ecosystem involved. Ideally such a conceptual model would be mathematical, but a word or pictorial model may be useful in predicting effects of environmental manipulations.

Five of the most characteristic features of pest management are outlined below.

1 The aim is management of the entire pest population, or a relatively large portion of it, rather than just localised infestations. The population to be managed is not contained within an individual farm, county, state or country, but is often international. International and national cooperation is thus essential.
2 The immediate objective is to lower the population density of the pest so that the frequency of population peaks, in space and time, above the economic threshold is reduced or eliminated.
3 The method, or combination of methods, of control is chosen to supplement the effects of natural control agents where possible. It is designed to give maximum reliable protection over a long term; minimum expenditure of effort and money, and the least objectionable effects on the ecosystem.
4 The significance lies in alleviation of the general problem over a long term and not just localised and temporary solutions. Harmful side effects are minimised or eliminated.
5 The philosophy is to manage the pest population rather than to attempt to eradicate it. The real significance of the concept concerns serious pest problems which defy solution through the more traditional approaches.

The term pest management implies some ecological understanding of the pest and in this lies its value. Decisions on the method used for control must be taken as soon as possible with the information available at the time. Further knowledge can be accumulated gradually to give fuller understanding which can then be used to modify the existing control method.

This approach to controlling pests is really quite new in concept and as yet is still more developed theoretically than in practice.

Pest eradication

In most cases pest control is undertaken to reduce the population density of the animal to a point at which the damage done is not of economic importance. Very rarely is complete eradication the goal, and even more seldom has it been achieved. Where complete eradication is attempted it is usually directed against pests of medical importance. Very successful campaigns have been carried out in many areas against diseases such as malaria, yellow-fever, dengue, etc. Complete eradication of agricultural pests is only attempted when a new pest, which is potentially very serious, has been introduced into a country and is still contained within a small area.

An eradication project was mounted in Florida after the Mediterranean fruit fly (medfly) was accidentally introduced. A total of 0·7 million hectares of citrus orchards and adjoining land was sprayed with insecticides, at a total cost of $7.9 million. However, the operation saved Florida's major agricultural industry which had (at that time) an annual gross value of about $180 million, the control measure in reality representing quite a small investment.

A veterinary pest on Curacao was very successfully eradicated in 1940 as a result of the first major trial of male-sterilisation (autocide). This was the screw worm (larvae of *Callitroga* sp.) on goats and after a programme lasting for 1 year total eradication of the pest was achieved (p.000).

Colorado beetle is accidentally imported in to the UK every few years but this important international pest of potatoes is usually destroyed quite rapidly. Since the 1933 Colorado Beetle Order most farmers and growers are aware of this pest and the danger it presents and infestations are usually reported promptly. Once a report is received the Ministry of Agriculture takes immediate action to destroy adults on the potato foliage and also to fumigate the soil in the infested area to kill any pupae that may be present.

However, the decision to attempt to eradicate an insect species is a grave biological responsibility. Eradication should not be attempted unless careful study of all aspects has produced convincing evidence that likely benefits more than balance the ecological impoverishment.

Further reading

Metcalf, R.L. and Luckman, W. (eds). (1975). *Introduction to pest management*. John Wiley and Sons: New York.

12 Physical methods

These methods use various forms of physical energy to remove or destroy pests and pathogens. They are mostly used for specific purposes and are not generally applicable to control pests and diseases of growing plants or crops in the field.

Use of physical factors

The use of lethal temperatures, both high and low, for insect pest destruction is of importance in some countries but of limited use in Africa. However, the use of cool storage in insulated stores for maize grain is practised locally in East Africa. The temperatures are not lethal and the method only reduces the metabolic rate of the insect pest and thus retards the rate of development. Kiln treatment of timber for control of timber pests is very widely practised in many countries in different parts of the world.

Temperate plant bulbs are often infested with mites, fly larvae (Syrphidae) or nematodes. Hot-water treatment (dipping) can be a very successful method of control if carefully carried out. Grain is dried to reduce moisture content and thus discourage most pests. In the Sudan cotton seed is heated to kill the aestivating larvae of pink bollworm (*Pectinophora gossypiella*). In Kenya hermetic storage of grain has been developed as a standard long-term storage method. The stores are Cyprus bins (about 70 of 1 million-bag capacity) which ensure that only a small quantity of air is enclosed within the sealed bin. The oxygen in the air is quickly used up by the respiration of the pests. Carbon dioxide accumulates quickly and results in the death of all pests and pathogens. Grain is often stored on the farm in butyl silos (Fig. 12.1). The addition of small quantities of diatomite fillers increases the effectiveness of this control. The abrasive effect removes the outer waxy covering of the epicuticle of the insects resulting in greater loss of water and thus increased dehydration; insecticide can penetrate the cuticle with greater ease.

Heating is a standard method for controlling many pathogens in certain situations where plants are not involved. Besides heat sterilisation of utensils, instruments, etc. by autoclaving (superheated steam under pressure, e.g. 120 °C at 1·03 kg cm² for 15 minutes) or using hot air (160 °C for 1 - 2 hours in an oven), nursery soil is often partially sterilised by heating. This is usually applied in the form of pressurised steam. Various types of apparatus are available, steam being most conveniently conveyed from a boiler through a flexible hose and applied through a perforated pipe which is buried in the soil. The soil to be treated should be loosely packed and friable, and covered with a tarpaulin or plastic sheet to retain the heat, which should be at least 80 °C for half an hour. Small quantities of soil may also be heated in ovens at 90 - 100 °C for one hour (dry heat).

Carefully controlled heat treatment may be used to eradicate some pathogens which are seed borne. Temperatures used are usually about 50 °C and the duration of the treatment about 1 hour, but temperature/time regimes have been worked out for some host/pathogen combinations (these usually can be found in definitive texts referring to particular crops). The material may be treated in a water bath in which the water is agitated and its temperature kept constant. Hot-air ovens have also been used but are more difficult to operate. Hot-water treatment is used to eradicate loose smut infections of cereal seed, *Ustilago nuda* on barley and *Ustilago tritici* on wheat; seed-borne black rot bacteria (*Xanthomonas campestris*) on cabbage seed; the bulb eelworm (*Ditylenchus dipsaci*) on bulbs and ratoon stunting disease of sugar cane setts (Fig. 12.2 on p. 68).

fig. 12.1 Groundnuts stored in a butyl silo in Nigeria

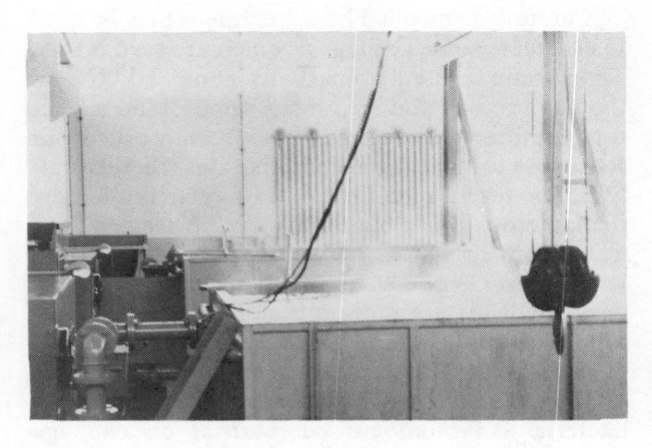

fig. 12.2 Hot water tank for treating sugar cane setts against ratoon stunting disease

Use of electromagnetic energy

The radio-frequency (long wavelength radiations) part of the spectrum has been studied extensively in the development of radio communications, radar, etc. It has been known for a long time that absorption of radio-frequency energy by biological material results in the heating of tissues. Control of insect pests by such heating is only practicable in enclosed spaces of small or moderate size (food stores, warehouses, timber stores). In a high frequency electric field different materials absorb radio-frequency energy to different extents. Some combinations of hosts and insect pests have favourable differential absorptions of energy and the insects can be killed without damaging the host material. Timber beetles in wood blocks have been killed in this manner but whether this treatment offers any real advantages over normal kiln treatment is doubtful.

Use of infrared radiation for heating purposes has not been developed sufficiently to be of great use in the field.

Various nocturnal insects, especially moths are attracted by ultraviolet radiation. Ultraviolet light traps have been successful in significantly lowering pest populations in some crops, but have failed when used on others. Many insects show distinct preferences for radiation of some visible wavelengths (i.e. certain colours).

Aphids and various fly pests show marked response to yellow colours. They will be caught in much larger numbers on yellow sticky-traps or in yellow water traps. Alternatively, a blue colour discourages aphid settling. A much lower infestation of cotton aphids has resulted from putting strips of blue material in the inter-row spaces between cotton plants in experiments.

The ionising radiations (X-rays, γ-rays) at low dosages sterilise insects but higher dosages are lethal. The use of these radiations in controlling pests of stored products, particularly of grain, is under investigation in a number of countries.

Further reading

Glass. E.H. (1975). *Integrated Pest Management: Rationale, Needs and Implementation*. Ent. Soc. Amer. (Special publication).

Part 3 Chemical methods of pest and disease control

13 The mode of action and presentation of toxic chemicals

Introduction

Pests and pathogens may be killed by exposing them to toxic substances. Highly toxic chemicals which are lethal to all or most forms of life are used in agriculture only as sterilants, particularly for the eradication of pests and pathogens from soil. More usually the substances used have a selective action. They are more lethal to the pests or pathogens which they are intended to kill than to the other plants, such as the crops they are designed to protect, or animals (including man).

Substances such as sulphur, and salts of copper or mercury, have been used for several centuries against agricultural pests and pathogens. The general use of chemicals to control the pests or diseases of agricultural crops did not occur until the late nineteenth century. At that time Millardet used Bordeaux mixture (copper(II) sulphate(VI) and lime) against downy mildew (*Plasmopara viticola*) on grapes in France. At the beginning of the twentieth century substances such as lime, sulphur, copper and mercury salts, and general sterilants such as methanal (formalin) were used increasingly to control pests and diseases of agricultural crops. From the 1930s and especially after World War II, a wide range of organic compounds was produced which have since revolutionised pest and disease control in agriculture and horticulture. The development, production and marketing of chemicals used for plant protection, is now a large and flourishing industry. These chemicals are generally called 'pesticides', they include fungicides and bacteriocides, used against plant diseases; insecticides and acaricides used against insect and mite pests; nematicides against nematodes; and herbicides against weeds. Some research is now aimed at developing chemicals which can be used against diseases caused by viruses and similar organisms.

Control of pests

Only rarely does chemical application kill the whole population. The individuals which survive may give rise to serious problems by the development of resistance. Chemical control usually has to be applied anew with each outbreak, it is, however, very quick in action. For the majority of pest outbreaks, chemical control is still the method by which the best results are obtained. The results of chemical control are more predictable than for other methods.

The different insecticides can be grouped according to their modes of action.

Repellants Designed to keep the pests away. They are usually employed against mosquitoes and other medical pests.

Fumigants Gases and smokes.

Stomach poisons These may be mixed with baits to encourage the insects to eat them or they may be applied to the leaves as a foliar spray.

Ephemeral contact poisons These are absorbed through the cuticle; they are applied as foliar sprays.

Residual poisons These remain active for a long period of time. They are applied to the leaves or the soil.

Systemic poisons They are absorbed and translocated by the plant; they are especially effective against sap-suckers. They may be applied as sprays or granules, to the soil or the foliage or trunk of the plant.

Antifeedants These chemicals inhibit feeding in insect and other pests. They were initially referred to as repellants, but this is a misnomer. They do not merely drive the insect away to another plant, they prevent the insect from feeding on that plant. In laboratory tests, insects, such as locusts and armyworms, have remained on treated plants indefinitely and eventually starved to death without eating the leaves. In field tests, the insects were free to wander elsewhere seeking food. They either found weed plants to feed on or died of predation and starvation. The earliest antifeedant used to protect plants was ZIP (a complex zinc salt). It was used to keep rodents and deer from feeding on the bark and twigs of trees in the winter. The first recognised antifeedant for use in insect pest control was introduced in 1959 by Cyanamid. Since then a number of compounds have been shown to possess antifeedant properties, but as yet a commercially successful antifeedant has not been found.

Control of diseases

Chemicals used to control plant diseases fall into three broad groups, according to their activity and how they are applied.

Sterilants, disinfectants or fumigants These have a wide range of activity and are often toxic to most forms of life (biocidal). They are generally applied to the soil before planting or sowing crops to eradicate pathogenic organisms. They can also be used for sterilising pots, boxes or buildings in which plants are to be grown and sometimes for the disinfection of plant produce after harvest.

Protectants These are usually applied to the seed or growing crop. They protect it from being infected by pathogenic organisms by forming a toxic barrier on the outside of the plant. These substances usually only affect fungi; they must not be phytotoxic (i.e. must not cause damage to the crop plant).

Therapeutants or eradicants These are applied to the seed or growing crop and may be very specific in action, as they can kill the pathogen after it has invaded the tissues of the host plant. Diseased plants may be cured in this way if the pathogen can be completely eradicated and the host allowed to recover.

Formulation

Commercially available pesticides usually consist of a mixture of the active ingredient (a.i.) plus other substances. These may include inert diluents, wetting agents, etc. which impart desirable physical and biological properties to the product. Inert additives are necessary to dilute the active ingredient so that the very small amounts required may be distributed evenly over the crop. They may also improve its storage and transportation. Emulsifiers, wetting agents, etc. are also needed to enable concentrates, wettable powders, etc. to be readily and evenly dispersed in water and to assist the adhesion of the chemical to the crop.

Many different formulations containing the same active ingredient may be used for several distinct purposes, because of the separate characteristics which the additives impart. Different types of formulation and their uses are listed.

1 Pure chemicals (unformulated) are rarely used now. Exceptions include volatile sterilants kept in pressurised containers, copper(II) sulphate(VI) (bluestone) and lime which are used to make Bordeaux mixture, and flowers of sulphur which can be used neat as a dust.

2 If the technical product is soluble in water, the pesticides can be prepared as a very **concentrated solution (c.s.)** which only requires dilution with water to the appropriate strength for spraying. The concentrated solutions usually have wetting agents or detergents added. This type of spray solution is typically homogeneous and spreads an even level of pesticide over the plant foliage.

3 Many solid substances that will not dissolve in water can be finely ground and formulated as **wettable powders (w.p.)**. These powders do not resist penetration of water, so they are miscible with water and readily form a suspension. **Water-dispersible powders** do resist penetration of water and remain as individual particles in suspension for a considerable period of time. Various additives, termed dispersants, can be included in the formulation of wettable powders and they will delay the process of sedimentation. Most fungicides are formulated as wettable powders with an inert material such as kaolin and a wetting agent.

4 **Dusts** consist of the active ingredient considerably diluted with an inert filler and ground to a very fine consistency. Seed dressings are specially prepared dusts containing higher concentrations of the active ingredient. They also contain 'stickers' which improve contact with the seed.

5 Oils and other water-immiscible liquids break up into tiny droplets when agitated with water. On standing the droplets coalesce rapidly to form a separate layer. This coalescence may be retarded or even prevented by the addition of a surfactant or emulsifier. With mixtures of oil and water, two types of emulsion may be formed. The oil may be dispersed as fine droplets suspended in water, which is then the continuous phase, giving an **oil-in-water (O/W) emulsion**. Alternatively, the water may be the disperse phase giving a **water-in-oil (W/O) emulsion**. The type of emulsion generally required for spraying crops is the O/W emulsion which, as water is in the continuous phase, may be diluted with water.

Pesticides that are insoluble in water may be dissolved in various organic solvents forming an **emulsifiable concentrate (e.c.)**. This can be diluted with water to an appropriate spray strength. Once on the plant, the water evaporates off. When most of the water has gone the emulsion 'breaks', i.e. the dispersed phase reforms and then can exert its toxic effect. Various chemicals can cause the 'inversion' of an emulsion which until then become useless for spray purposes. Sometimes emulsions are caused accidentally to 'cream' (named from the analagous creaming of milk), resulting from differences in specific gravity between the dispersed and continuous phases of the emulsion. Some oil formulations of pesticides are used neat, without emulsification in water, for ULV applicators. The oil acts as a carrier for the very small droplets of spray. In hot climates, water evaporates too quickly so that if an emulsion is used the toxin can fail to reach the target.

6 Some pesticides can be more suitably formulated as **miscible liquids (m.1.)**. In this case the technical product is usually a liquid and is mixed or dissolved in an organic solvent which is then diluted and dissolved in the water carrier.

7 A relatively new method of insecticide and some nematicide formulation is granulation. **Granules** are small solid particles 2–5 mm in diameter. They are becoming widely used to apply systemic organophosphorus insecticides for the treatment of seedling crops. The main advantage of granular formations is that the insecticide can be placed so that it gives maximum protection to the plant with minimal danger of large-scale soil pollution. Granules also reduce the risks run by the operator. This is of particular importance with highly toxic chemicals. In granular formulations the active ingredient is less affected by the soil than if applied in one of the liquid formulations. Many pesticides are strongly absorbed by soil particles and rapidly become ineffective once they reach the soil. The rate at which the pesticide escapes from granular formulations is mainly controlled through the rate of leaching by rainwater. However, the organophosphorus compounds formulated in granules are generally insoluble in water. Other major factors controlling rate of pesticide release from granules are temperature, dosage and size of granule. The 5 major organophosphorus insecticides at present formulated as granules are dimethoate, phorate, disulfoton, chlorfenvinphos and diazinon, used mostly against fly maggots, beetle larvae, aphids and some nematodes.

Nematicides are commonly formulated as granules to facilate application to the soil. Granules are sometimes applied broadcast, but more typically they are applied by the bow-wave technique to the rows of seeds as they are sown. Where labour permits, they may be given as spot applications round the bases of individual plants, using hand applicators. 'Rogor' and 'Birlane' applicators are generally available for this purpose. The body of the granule is made of various inert substances. For soil application both phorate and disulfoton granules are made of Fuller's earth. Fine granules, often less than 2 mm diameter, called prill, may be formulated for special application techniques to soil.

Granular formulations made for foliar application are sometimes advantageous to use. These granules are made of finely ground pumice, which although rather expensive to produce, are very effective because the formulation is sticky and lodges easily in the plant foliage.

8 The most recent development in pesticide formulation is the new technique of **micro-encapsulation**. Research at Rothamsted Experimental Station, UK, has shown that it is possible to encapsulate an insecticide in a non-volatile

envelope of cross-linked gelatin. It is then non-toxic by contact but is toxic to insects ingesting it. The formulation can be given considerable resistance to weathering by the addition of suitable stickers. The capsules are so tiny that the formulation looks like a slightly coarse powder.

This type of formulation can be expected to be of great promise for the control of leaf-eating insects in the Tropics. The formulation has the advantage of being much safer to handle than other pesticide formulations where dangerous chemicals are being used. It presents far fewer hazards to beneficial insects such as predators, parasites and pollinators. There have been recent reports of bees collecting the capsules in mistake for pollen grains which are about the same size, this is known to have caused some larvae deaths in the hive.

When a contact kill is required it is possible to prepare leaking capsules which will release the poison over a fixed period of time. At Rothamsted it was found that under warm conditions a standard wettable powder of DDT lost over 90 per cent from the target area in 35 days, whereas a leaking capsule formulation lost only about 20 per cent over the same period of time.

9 Other special purpose formulations include **liquid seed dressings** and **fungicidal paints** used to protect wounds and **pelleted seeds.**

The rate at which a formulated product is used depends upon the amount of active ingredient it contains. The formulation of a pesticide may dictate special precautions in the way in which it is applied. This information is printed on the labels of commercially available products and these should always be read carefully and the instructions followed. Any special queries should always be referred to a competent plant pathologist or entomologist.

Spray additives

Spreaders, wetters or surfactants Substances added to the spray to reduce the surface tension of the droplets so as to facilitate contact between spray and sprayed surface. If water without a spreader falls on a waxy leaf, it will collect in large drops and will then run off leaving the leaf surface dry.

The incorporation of a spreader is not standard production practice in the manufacture of most modern pesticides. For crops with particularly waxy leaves (such as *Brassica* spp.) or against pests with particularly waxy cuticles (like mealybugs and woolly aphids) it is necessary to add extra wetter to the spray solution. Sometimes, when extensive run-off is required to enable the pesticide to penetrate to the lower parts of a dense crop or on to the soil surface, this can be achieved by addition of extra spreader to the spray. From a physical point of view wetting and spreading are not quite the same, but for practical purposes they can be regarded as synonymous.

Spreaders and surfactants exist in 3 forms, classified according to their ionising properties: non-ionic, anionic and cationic. Surfactants are defined as surface-active components.

The non-ionic detergents depend upon a balance between hydrophilic and lipophilic properties throughout the molecule for their wetting properties. The advantages of these substances include the fact that they are incapable of reacting with cations or anions of other spray components or in hard water. They are not hydrolysed in either acid or alkaline solutions so they are quite stable. However, phytotoxicity of supplements has to be taken into account for it can be serious.

Anionic spreaders include soap, sulphonated alcohols and sulphonated hydrocarbons.

Cationic spreaders include the quaternary ammonium and pyridinium salts. The main advantage of cationic spreaders is that they do not react with ions of heavy metals.

The incompatibility of anionic and cationic additives must be remembered if it is necessary to add several supplements to a spray mixture.

Dispersants (protective colloid) These delay sedimentation in suspensions. Sedimentation is always a danger with suspensions and sprays must be of uniform concentration to ensure even application. The methyl celluloses and the sodium carboxymethyl celluloses are colloids which have proved effective as dispersants.

Emulsifiers These modify the properties of the interface between the disperse and the continuous phases in emulsions. Many of the spreaders and surfactants may also function as emulsifiers (e.g. soap).

Penetrants Oils may be added to a spray to enable it to penetrate through the waxy epicuticle of an insect pest more effectively. Some of the most effective sprays against locust swarms have been solutions of dieldrin in light petroleum oils.

Humectants Substances added to a spray to delay evaporation of the water carrier. Glycerol and various glycols are commonly used. Humectants are more frequently used with herbicides than with insecticides.

Stickers Substances including methyl cellulose, gelatine, various oils and gums, used to improve the tenacity of a spray residue on the leaves of the crop. The need for maximum spray retention is particularly important in the Tropics where rainfall is often torrential. Generally a deposit of fine particles is more tenacious than one of coarse particles. Stickers usually enhance the tenacity of a deposit and its retention on the plant, although they may retain their wetting properties also and thus cause the deposit to be washed off by prolonged rain or dew. Some stickers break down on drying and form insoluble derivatives; these can, of course, greatly enhance retention of the deposit.

Lacquers In order to achieve a slow gradual release of insecticide in certain locations it is possible to formulate some insecticides into lacquer, varnish or paint. The painted area then releases the insecticide slowly over a lengthy period of time. The insecticides used in this manner have been mainly organochlorines and in particular DDT and dieldrin. Incorporation of insecticides into paint is of some value but the lacquers tend to remain after the pesticide has dispersed which limits their use. This practice is generally more suitable against household and stored products pests than crop pests.

Synergists Substances which cause a particular pesticide to have an enhanced killing power. These are sometimes called **activators**. The way in which synergists act is not always fully understood. Some are known to operate on a biochemical level where they inhibit enzyme systems which would otherwise destroy the toxicant. Usually the synergist itself is not insecticidal. Piperonyl butoxide is a synergist for the pyrethrins and certain carbamate insecticides. Other commonly used synergists include piprotal, propyl isome, sesamin and sesamex. Many of these are particularly effective on the pyrethrin group of insecticides. Some pairs of organophosphorus insecticides have a mutually synergistic action (sometimes called **potentiation**).

Persistence

In temperate countries much is known about the amount of pesticide reaching the target area and how long it persists there. The relationship between the amount of pesticide present and the level of control of the pest obtained is thus fairly well established.

Relatively little work has been done to study the behaviour of pesticides under tropical conditions. The variety of crops and ways in which they are grown make it difficult to draw general conclusions. Data on this subject are scanty and control measures largely speculative.

There are also application problems in some tropical countries arising from local agronomic methods and environmental conditions. Some of these arise from the practice, common among peasant farmers, of growing crops such as coffee and cotton in small patches. This makes it impossible or uneconomic to use any but the smallest hand applicators. Even with crops such as coffee, grown in larger plantations, the variation in plant growth and pruning methods often makes it difficult to spray all parts of the plant effectively. Apart from the difficulties of ensuring that sufficient quantity of the pesticide reaches the target area, differences in the persistence of the deposits will be expected.

Temperature In the Tropics plant surfaces can attain temperatures at which no non-systemic insecticide applied in commonly used formulations would persist for more than a few days. This has been shown at Rothamsted Experimental Station in the UK, by experiments with insolated surfaces. Under tropical conditions the majority of even the most persistent insecticides may well be lost from the target surface within a few days. For example, dieldrin deposits lost 25 per cent at 20 °C and 95 per cent at 40 °C, in 24 hours, at the same low wind-speeds (3 km per hour). However, when the deposit level is very low (about 0·005 μg per square centimetre on a glass surface, or alternatively 0·01 – 0·02 μg per square centimetre on a cotton leaf) it no longer obeys these rules, but remains very firmly bound to the surface for a very long period. These data concern insecticide applied as sprays of wettable powders, emulsions or dusts.

Most agricultural crops in the Tropics are exposed to the sun and so rapid breakdown of insecticide can be expected, although for some plantation crops the trees may form a canopy with shade underneath. Dieldrin applied to tree trunks and the underneath of branches for tsetse fly control will usually persist, since it is shaded from the direct sunlight, and can effectively kill tsetse resting there for periods of up to 5 months.

Rain Rain is often torrential in the Tropics and thus becomes a serious hazard to the retention of chemicals by plant surfaces.

Work with protectant fungicides on cocoa and coffee has shown that many of these substances may be completely removed after less than 150 mm of rain. As most plant diseases are very active during wet seasons, the maintenance of good fungicidal cover during this period is very important. Flexible spraying schedules, depending on rainfall rather than fixed time intervals, are more efficient. However, the movement of fungicide in rainwater can be of use. Rain causes redistribution of fungicide throughout the foliage canopy. Overhead spraying of perennial crops such as coffee and citrus has been particularly beneficial in controlling some diseases.

Unfortunately, however, not a great deal of research into the persistence of most pesticides has been carried out in tropical countries and often spray programmes tend to be too speculative.

Nomenclature of pesticides

Nomenclature of pesticides is confused, as there are usually several names which refer to one substance. Names can be grouped as follows:

1 The chemical name describes the structure and composition of the chemicals involved. This may be straightforward for fairly simple compounds, but with complex organic compounds different chemical names may be used to describe the same compound according to usages in different countries. This could be remedied if all countries followed IUPAC recommendations.
2 The common name may be identical to the chemical name with simple compounds, or it may be an abbreviated form of the chemical name when this is complex. Generally, most chemicals are properly referred to by their common names which are approved by the International Standards Organisation, the British Standards Institution or the American National Standards Institute.
3 The trade names under which the different formulations of the same compound are marketed vary widely. A trade name is characteristic of the chemical company which produced the product. Trade names are a particular source of confusion, as they may become generally, but wrongly, accepted as a common name where one company has a marketing monopoly of a particular chemical in one area.

All products used to control pests or diseases should clearly state the common (standard) name of the chemicals they contain and the amount (a.i.) in the formulated product. In this book, common (standard) names are used. They are supplemented with chemical names where this is relevant and by trade names in parenthesis where necessary.

14 Methods of application

The successful control of pests and diseases by chemicals often depends as much on the way in which pesticides are applied to soil or plants as on the substance used. Efficient control depends upon correct application. The chief points to be considered when pesticides are applied are listed below.

Dosage The correct dosage is the amount of material needed to give adequate control. Use of excessive quantities is wasteful and uneconomic. It may also damage the crop plant or create an environmental hazard. Too small a dosage may not give adequate control and may encourage the development of resistance.

Target This is often the site of the pest or pathogen. In order to be effective some chemicals, e.g. non-systemic pesticides, must come into direct contact with these. Soil sterilants must be thoroughly and evenly distributed throughout the soil. Protective fungicides must be adequately distributed over the susceptible parts of the crop. Systemic pesticides which are absorbed by the roots of the growing crop plants must be placed in the rooting zone.

Time of application Pesticides must be applied when they are most beneficial in controlling a pest or pathogen and this is before the damage is done. Pesticides are often applied as a routine measure to ensure that damage does not occur. To wait until the first signs of an attack by a pest or pathogen are apparent before taking action can be a very dangerous procedure for some attacks can develop rapidly into dangerous epidemics. The frequency of application must be calculated to ensure continued protection while the crop is at risk. Pesticides may be rapidly removed by weathering, and continuing growth produces a larger crop area to be protected.

Physical placement Various techniques and machinery have been devised to put the pesticide on the target at the correct dosage. Machinery must always be properly adjusted and kept in good working order. The appropriate pesticide formulation and technique should be matched to the machine that is used.

External hazards Many pesticides are toxic even if only mildly so and care must be observed when using them. They may present risks to the operator as well as to crops and the rest of the environment if improperly used. The precautions written on the product label must always be followed.

There are many examples of poor pest or disease control and of damage to crops, livestock and man which have resulted solely from the incorrect or careless application of pesticides. Whereas, if they had been correctly applied they would have done a safe and efficient job.

Application to the soil

Many pests or pathogens of crop plants are soil-inhabiting organisms. Seedlings are often killed by attacks of soil-borne pathogens and it is economical to control these in seed beds, nurseries and other small-scale, intensive cultivations by treating the soil. Pesticides may also be applied to the soil on a field scale.

Pests and pathogens in soil are often controlled by the use of general sterilants. These can only be used before crops are planted. The amount of soil treated is limited by the procedure required. Frequently, the technique used only partially sterilises the soil, but succeeds because harmful organisms are killed before beneficial ones. Soil may be sterilised by heat (p. 67).

Chemical sterilants are usually applied to soil either as drenches or by injections of highly volatile materials into the soil but some are now available in granular formulations. The injections are applied at a specific depth and specified intervals depending upon the dosage and the material being used. All sterilant chemicals applied to soil have some fumigant action so that they can penetrate minute spaces in the soil. Covering the soil with tarpaulin or plastic sheeting helps to retain the fumigant vapour. The process is most efficient when soil is loosely packed, friable, slightly moist and not too cold. Lumpy, wet, cold soils hinder the movement of the chemical through the soil and delay its dispersal after treatment.

Time must elapse to allow dispersal of phytotoxic chemicals before sowing can commence. One way to test whether soil is safe, is to sow cress seeds, or some other sensitive, quickly-growing seed, on to samples of the soil at various intervals. The soil is safe when germination and subsequent growth of the test crop is normal.

fig. 14.1 Manually operated soil injector

Hand-operated injectors (Fig. 14.1) can be used to apply these chemicals to small areas but, large-scale, mechanically driven soil-injection equipment is used to apply nematicides over large areas. This type of operation is usually economic only on a few high-value cash crops such as tobacco and some vegetables and fruits. A simple way of applying pesticides to soil is to incorporate the chemicals during normal soil-cultivation procedures. Powder or granules can be applied by rotavating them in, or applying them as a band in or by the side of the seed drill. Substances applied in this way must not of course be phytotoxic to the crop. Sterilant chemicals can only be applied to fallow soil, although some fumigants may be injected into the soil beneath perennial crops, e.g. for controlling banana nematodes.

Application to seed

Pesticides applied to seeds may serve several purposes. They may eradicate seed-borne pests and diseases and control soil-borne pests and diseases which may attack the growing seedling. If the pesticide is systemic, it may be absorbed by the roots of the growing plant to protect the shoot from pests or pathogens above soil level.

The vital importance of clean seed was stressed in Chapter 8. Few chemicals effectively eradicate diseases borne internally in seed material, but carefully controlled heat treatment may be effective against some pathogens (p. 67).

Seeds are less susceptible to phytotoxic chemicals than are the growing shoots of plants. Some chemicals which are phytotoxic to foliage can therefore be used as seed dressings where they are more easily tolerated. The earliest form of seed dressing used was to steep the seeds in liquids such a brine or wine. The dressing will form a protective zone around the seed, and the extent of the zone will depend upon whether the pesticide has any fumigant or systemic action. In the past, seed dressings have been used mainly against smuts and other seed-borne diseases, but now there are many insecticides which can be successfully formulated as seed dressings. For instance, those used to combat wireworms, chafer larvae and shoot flies. Preparations of systemic compounds can be used to protect against aphids and other sap-sucking insects on the young plants, and to protect young plants against soil-borne diseases and early attacks of airborne pathogens.

Seed dressings can be applied as liquids which are adsorbed on to the seed coat. Some powders are sufficiently adhesive to stick directly on to the seed coat, others have to be stuck on to the seed with the aid of a sticker, e.g. kerosene or methyl cellulose. Only a small quantity of pesticide is needed to treat a relatively large amount of seed. Thorough mixing

fig. 14.2 Hand seed dresser

fig. 14.3 'Rotostat' seed treater

of the chemical formulation and seed is needed to ensure even distribution. This is usually done in a rotating drum with baffles which may be driven by hand or machine (Fig. 14.2). Dressing seeds with toxic chemicals in dust form is extremely hazardous and operators should always wear goggles and a respirator. Liquid seed dressings avoid this problem. There are modern methods of continuous application where the dressing is applied as a fine mist, through which the seeds fall at a controlled rate. The rotostat seed dresser (Fig. 14.3) can handle a range of different formulations and achieve efficient seed coverage by a rotary stirring action.

Organomercurial compounds, used extensively against a wide range of seed-borne diseases, are now being replaced by less toxic fungicides such as thiram and the recently developed systemic fungicides. Chemicals are most effective against pathogens carried on the outside of seeds but organomercurials are powerful eradicants even of deep-seated infections. This is enhanced by the slight fumigant action of some preparations.

New developments in systemic fungicides are constantly improving the chemical control of internal seed infections. Soaking seeds in a solution or suspension of fungicide for a certain length of time may be effective in controlling deep-seated infections but this method is applicable only to relatively small quantities of seed. Vegetative planting material is often dipped in a fungicidal suspension or solution before planting to control soil-borne diseases. For example, in control of pineapple disease (*Ceratocystis paradoxa*), fusarium sett rot, etc. on sugar cane planting material.

The development of precision seed drilling gave rise to pelleted seed. Pelleting is a special technique which envelops the seed with a smooth, round protective coat. The coat consists of several chemicals, usually fertilisers and pesticides, together with a cementing material, usually a type of cellulose. This coat, which is a form of seed dressing, gradually dissolves in the soil moisture when the seed is planted. The chemicals are released into the rooting zone of the young seedling. Irregular-shaped seeds are given a smooth, spherical shape by pelleting and very small seeds are made larger and are more easily handled. The method is economically feasible only on a fairly high-value crop and for small quantities of seed, as it is expensive.

One disadvantage is that under drought conditions pelleted seed suffers from impaired germination. Despite higher cost, however, an ever-increasing range of pelleted seeds are being supplied by the major seed companies.

Application to crops

Pesticides are most frequently applied to growing crops, usually in the form of liquid sprays. Dusting is used to a limited extent and paints may be applied to localised areas of perennial crops for special purposes (e.g. panel diseases of rubber). Fumigation is used for the protection of stored crops and to a more limited extent under field conditions. Smoke generators are a minor method of pesticide application. Aerosols can be used for the small-scale application of pesticides.

Spraying methods

There are several different techniques for spraying pesticides which vary according to the volume of solution applied to the crop. A variety of spraying machines have been designed to deliver the different volumes.

High-volume spraying The term 'high volume' usually applies to spray rates of more than 400 litres per hectare. With high-volume spraying of pesticide the carrier is always water, and the usual quantity involved is in the region of 600-1 200 litres per hectare. If a run-off spray from the upper parts of the plants on to the lower parts or on to the soil is required, then the water volume may be doubled up to the extent of 2 400 litres per hectare. Hydraulic sprayers which pump the liquid directly through spray nozzles are usually used for high-volume spraying.

High-volume spraying has several serious disadvantages, the first being the problems involved in obtaining and transporting the large quantities of water required; this is especially difficult in areas where piped water is not available. In many parts of the drier Tropics obtaining sufficient water for this purpose can be a very difficult problem. The cost of high-volume spraying equipment is usually considerable. The equipment is so bulky that its operation often requires a large tractor.

Because of these problems alternative methods of application for the treatment of large areas have been sought. For instance, a jet of liquid can be dispersed into fine droplets by the force of a copious air-flow. This method has been successful for the dispersal of DDT/petroleum mixtures from aeroplanes for locust control, where the air speed alone is sufficient to break up the jet of solution into droplets, which are then dispersed by the slip-stream. A helicopter can be even more effective for this method because the down-draught from the rotor blades is very strong. For use on the ground the air-stream can be provided by a fan mounted horizontally, or alternatively by a turbine fan. These sprayers using energy provided by the air-stream are called mist blowers or air-assisted sprayers and the technique is called atomisation.

Typically, the droplets produced through atomisation are much smaller than those produced by hydraulic sprayers. Because of the smaller droplet size, smaller quantities of spray are needed per hectare and the use of organic solvents such as kerosene, petroleum oil, or fuel oil in place of water is quite economically feasible. If water is used as carrier the concentration of active ingredient can be increased, this is known as **low-volume spraying**.

Low-volume spraying This term is usually applied to the application of liquid volumes in the region of 5-400 litres per hectare. However, the typical rates for ground application are commonly 100 - 200 litres per hectare, whereas for aerial application they are around 15 - 75 litres per hectare. Mist blowers or air-blast machines are typically used for ground-based low-volume spraying. The carrier liquids may be organic solvents, although water is frequently used. Mist blowers can be used to spray by two different methods, either as blast spraying or drift spraying.

Blast spraying is usually carried out in orchards, or plantations of such crops as apples, mangoes, coffee or rubber. The air all round each tree is replaced by a mixture of air and spray droplets, and generally a good even cover is achieved using as little as 90 litres per hectare. Run-off of pesticide is eliminated. The small droplets dry out quickly and the chemical becomes firmly attached to the foliage.

Drift-spraying relies upon the movement of ambient air to carry the pesticide mist through the crop. Its use is dependent upon favourable weather conditions. The optimum size of droplet is 80 to 120 μm. Very fine droplets either evaporate or are lost in the wind and never settle, larger droplets seem to settle too rapidly.

Ultra-low-volume (ULV) spraying The ULV application technique was developed in East Africa in the control of the desert locust, shortly after World War II. The solutions used were DNOC and dieldrin in diesel oil. Development of ULV for crop spraying started much later. The technique consists essentially of the production of very small droplets (c.70 μm diameter) carried in light oil. These can be delivered aerially or from the ground. The droplets are either blown by a fan or allowed to settle naturally on the crop (drift spraying). Experience with ULV techniques has shown that the size of the droplets needs to be precisely controlled in order to achieve accurate cover of the target area. On p. 89 the topic of controlled droplet application (c.d.a.) is considered in more detail.

Ultra-low-volume spraying has been developed to overcome the difficulties of spraying crops in areas where low rainfall causes water shortage. The physical problems involved in handling large quantities of water; increased chemical and labour costs and consideration of environmental pollution, have all encouraged development of ULV techniques.

In the early 1960s much of the development work on ULV for crop spraying was done by the Plant Pest Control Division of the United States Department of Agriculture (USDA) in cooperation with American Cyanamid. When applying ULV sprays the pilot generally flew the spray plane rather higher (3 - 6 m) than in conventional spray application, where the altitude is usually about 1 - 2 m. The higher altitude increases the spray swath by about 3 times. Generally, with the Cyanamid ULV method the spray swath was about 30 m compared to the more conventional 12 m. Aerial spraying can also be carried out using hydraulic nozzles (Fig. 14.4 on p. 80).

The real establishment of ULV techniques for crop spraying was probably in 1963 with the good results obtained by malathion for the control of insects on cotton. From that time a large number of pests have been successfully controlled on a wide range of crops.

It is usual now to apply the spray from a height of 2 - 4 m using swath widths of 15 - 25 m, depending on type of aircraft and equipment used. The operation takes less than half the time of conventional aircraft spraying because of the small amount of liquid being sprayed.

In the UK an improved system of control over agricultural aviation was introduced by the Civil Aviation Authority in 1974, and now aerial spraying operators may select pesticides for application from the air only from a 'permitted list' compiled by the Ministry of Agriculture, Fisheries and Food (MAFF) under the Pesticides Safety Precautions Scheme. The permitted list specifies the formulations of the various pesticides that are allowed to be used.

ULV ground-spraying was introduced quite recently. A great variety of equipment is used now and this restricts the development of one general technique. The advantages of changing from conventional methods to ULV for ground-spraying are less apparent than with aerial spraying. It is useful where transport of water is a problem, as in **ultra-low dosage spraying** where only small quantities of pesticide are needed.

ULV sprayers require fine droplets to ensure

fig. 14.4 Aerial spraying of citrus: fixed wing aircraft equipped with a boom of hydraulic nozzles

adequate cover. If the droplets are too fine, losses due to drift and evaporation become too great. Recent development of rotary atomisers has ensured the consistent production of droplets in a narrow range of sizes. Several manufacturers now produce ground-based ULV spraying equipment.

Generally, ULV application of fungicides for control of diseases has had only limited success. This may be because of the difficulty of getting a reasonably large quantity of chemical sufficiently well dispersed throughout the foliage. However, aerial application of fairly low volumes 10 - 20 litres per hectare of concentrated fungicidal suspensions have shown reasonable success for specific diseases; for example, Sigatoka disease of bananas. Hand-directed ULV spraying has been successfully used to control vegetable diseases.

Unlike pests, pathogens are not generally actively mobile and do not pick up pesticides as they move. A fungicide must therefore reach the site of potential infection or sporulation of the pathogen if it is to be effective. Rainfall does help considerably to redistribute fungicide throughout plant canopies. The development of systemic fungicides has to some extent modified these requirements. Techniques which did not work well with the old surface protectants now show promise with fungicides which move within the crop plant.

Droplet size The major objective in crop spraying is to spread the active chemical evenly over all the plant surfaces so that a lethal dose is available for pest contact. When large spray volumes are used the coverage is often far from complete because of the coalescence of drops and subsequent run-off.

Thus the pesticide should be distributed over the plant surface in spray droplets that are as small as possible to produce a complete coverage. The droplet-density required in a given spray operation depends upon various factors: type of pesticide (fungicide or insecticide); mobility of pest to be controlled; mode of action of insecticide (systemic, contact or stomach poison). Thus when a contact insecticide is sprayed against a sluggish pest, a higher droplet-density is required than for the spraying of a stomach poison for the control of a more mobile insect. In the case of fungicides an even better coverage is required for effective control.

Generally, it would be expected that the smallest droplet size would be the most effective, but very small droplets do not fall freely. They tend to be carried by wind and air currents. It appears that droplets of less than 30 μm become permanently airborne. They will be carried by the surrounding air so that they will not touch large target surfaces such as cotton leaves at all, but they may settle on small-diameter targets such as the needle leaves of conifers, hairs of caterpillars, etc. Thus for control of adult mosquitoes the optimum droplet size is from 5-25 μm, but for tsetse in vegetation it is 10-30 μm.

For spraying agricultural crops the situation is complicated — for many cotton pests it has been found that a droplet size of 20-50 μm was most effective. Many crop areas are relatively small and are often situated near other crops which means that the spray must be deposited in a rather limited target area. Consequently, optimum droplet size in many crop-spraying programmes may be decided upon mainly by the necessity to avoid spray drift, and thus larger droplet sizes will often be used.

In aerial crop spraying for insect control the optimum droplet size appears to be in the range of 80-120 μm. The smaller droplet sizes are more suitable for the treatment of large areas. In ground crop-spraying operations the ULV droplet size will usually be in the range of 60-90 μm, although with systemic insecticide the optimum droplet size may be larger.

Solvents Smaller droplets have a relatively larger surface area, which implies that the rate of evaporation of carrier liquid spray droplets is higher with smaller droplets. Consequently, solvents used in ULV formulations must have low evaporation rates, and water is seldom suitable. The use of the more volatile carrier liquids could also lead to evaporation of the solvent in the atomiser and may cause crystallisation of the pesticide. Liquid pesticides are sometimes sprayed undiluted in ULV application, but more often require some solvent.

The solvent must be non-phytotoxic and of low volatility and viscosity. It should be compatible with the pesticide and be capable of dissolving it readily. Not many of the solvents generally available will fit all of these categories, so good ULV solvents are not easy to find. 1,2-Dimethylbenzene (xylene) is too volatile, and other aromatic hydrocarbons of lower volatility are often highly phytotoxic. Alcohols and ketones show higher phytotoxic effects at lower volatility, and most other common solvents show a similar variation in properties. Sometimes, a mixture of various solvents can be employed with considerable success. **Adjuvants** which dramatically decrease the phytotoxicity of many solvents of low volatility have been developed by Phillips-Duphar. Thus the range of solvents that can now be used for ULV formulations has increased.

Conventional emulsifiable concentrates usually have a low flash-point, but this seldom represents a serious fire hazard since the concentrate is mixed with a large volume of water before spraying. However, with ULV techniques the solvent must of necessity be of low volatility, for when using rotary atomisers with electrical systems and many rotating parts the possibility of electrical discharge is always present.

These adjuvants enable a wider range of solvents with low volatility to be used for ULV formulations and many of these formulations are now termed **special ultra-low-volume (SULV) formulations**. Their characteristics are high concentration, low volatility, low phytotoxicity; with a much lower viscosity than most ULV preparations, and with a flash-point above 75 °C.

Spray residues In conventional spraying the spray liquid typically consists of a large amount of water, containing various wetting agents, dispersants and/or emulsifiers. The pesticide is generally present in a finely dispersed phase, liquid, when using emulsions or solid, in the case of wettable powders. In ULV sprays the pesticide is generally present as a true solution in an oil carrier, or sometimes the technical material is sprayed without dilution. Such differences in the spray make-up will affect the behaviour of the spray droplets on the biological target.

After deposition of a spray droplet on a leaf, the droplet will assume a particular physical shape and will spread over the leaf surface. With conventional aqueous spray liquids this spreading depends greatly on the physico-chemical properties of the leaf. On hydrophilic leaves the droplet will spread to a thin film, but on lipophilic leaves the droplets tend to retain a more spherical shape. The spreading of oily liquids on most smooth leaf surfaces is much better than that of aqueous solutions. The SULV formulations spread to a thin film on most leaves, even lipophilic leaves, with very little run-off. However, if run-off is required ULV application is not the most suitable method of application; high-volume spraying is usually considered the most appropriate.

The formation of a residue from an emulsion droplet is a very complicated procedure, as evaporation of the water, breaking of the emulsion, and crystallisation of the pesticide from the oil phase can occur simultaneously. Often following use of ULV sprays the crystalline residue on the leaves is particularly coherent and very resistant to dislocation. ULV residues seem to resist the dislodging effects of rain better than the more conventional formulations.

Dusting

There are times when it is more convenient to use a dust on a crop instead of a spray. The need for water is obviated; the dust may be bought ready for use and is more easy to handle than spray concentrate.

For dusting the active ingredient is typically diluted with a very finely divided 'carrier' powder, such as talc. The dust is usually applied by introduction into the air stream of a fan or turbine blower. Some problems may arise in practice. Frequently the powder 'cakes', usually through absorption of atmospheric moisture, or 'balls' in the hopper, through static electrification. Also, it is quite difficult to ensure that the dust is homogeneously mixed with the pesticide.

It has generally been found that dusting is only practicable during the calmest weather, and that the best results are obtained when the dust is applied to wet or dew-covered plants. When dusts are applied to dry foliage often not more than 10–15 per cent of the applied material sticks to the foliage. Thus there are not many occasions in the Tropics when dusting is a more suitable application method than spraying, except in some dry regions.

Dusting machines are usually simpler than spraying machines and have the advantage of not relying on large volumes of water; they are therefore much lighter to use. Water for spraying is a limited factor in many areas, particularly as it may have to be transported long distances over difficult terrain. Dusting has been used with some success against blister blight of tea in India and against coffee rust in Brazil, but is usually less efficient in disease control as the pesticide is more easily removed by rain. A new technique called **wet dusting** has been developed. A fine mist of water is delivered to the crop at a very low rate as it is dusted. This is claimed to overcome some of the disadvantages of simple dusting.

Fumigation

The toxicity of a gas to a pest is proportional to both its concentration and to the total time of exposure against that pest. Research into the properties of gases has shown that usually fumigation is only successful in completely enclosed spaces or with special precautions to lengthen the time of exposure. Stored products can be fumigated in special chambers or under large gas-proof sheets. Some field crops are treated by drag sheets in which the fumigant is enclosed below a light impervious sheet dragged at a rate dependent on its length behind the applicator.

Smoke generators

Smoke generators contain a combination of

pesticides and a combustible mixture which burns in a self-sustained reaction at a low temperature so that the minimum amount of pesticide is destroyed during volatilisation. The combustion reaction is generally rapid and lasts for about 10 - 20 seconds.

Aerosols

Aerosols contain the toxicant dissolved in an inert liquid which is gaseous at ordinary temperatures but liquifiable under pressure. When the pressure is released the solution is discharged through a fine nozzle and the solvent evaporates. The toxicant is dispersed in a very finely-divided state. Chloromethane, at 5·6 kg per square centimetre and dichlorodifluoromethane (Freon), at 6·3 kg per square centimetre at ordinary temperatures are two widely-used aerosol solvents. These solvents are, however, phytotoxic and so aerosols are generally used only against medical, veterinary or household pests. In 1976, Bayer introducted a water-based aerosol formulation for use on pot plants and ornamentals.

15 Equipment for application to standing crops

All spraying systems consist basically of a tank for holding the spray liquid and a mechanism for producing the spray droplets. There are three fundamentally different mechanisms for doing this.

1 **Hydraulic sprayers** in which the spray liquid is forced through a fine nozzle under pressure. The pressure may be applied directly to the liquid by a hydraulic pump (pump systems) or indirectly by compressing air into the tank (compression system). Most hand-sprayers work on the hydraulic principle and this mechanism is also used in most high-volume application systems.
2 **Air-assisted sprayers, atomisers or mist blowers**, in which the spray liquid is drawn through a simple tubular nozzle and the spray produced by the shearing force of a high-velocity air stream across the nozzle. Some simple hand-atomisers can be used for treating individual plants, but those used on crops are all power operated. They are usually used for medium- or low-volume application.
3 **Ultra-low-volume (ULV) sprayers**, where the spray droplets are produced by the centrifugal force of a spinning disc.

The general function of all spraying machines is to produce an even cover of pesticide over the plant using as little liquid as possible. Two components of the machines are of particular importance to ensure this function: the filters and nozzles. They must be checked frequently.

Filters All spraying machines are equipped with a series of filters to ensure that no coarse particles are permitted to pass into the feed pipe which would block the nozzles. Filters are vitally important, particularly since the mixing of the spray and the filling of the tank often takes place in the field. Without the main tank filter the spray would frequently become contaminated with insect bodies, leaves, pieces of grass and other detritus, which would clog the nozzles and prevent effective spraying. At various points in the system of pipes, additional filters may be placed and the nozzles themselves may also be fitted with gauze filters in various positions according to their particular design.

Nozzles These are of vital importance in hydraulic sprayers as they are the detachable apertures which break up the pesticide liquid into the spray. Many different types of nozzles are made, and each type gives its own particular spray pattern. The nozzle disc controls the final shape of the spray pattern. In **cone nozzles**, which have a spray pattern in the form of a hollow cone, the disc is perforated by a series of small holes arranged in a circle. In **fan nozzles** the perforations are in an elongated horizontal slit. Nozzle discs are often made either of tungsten steel or more recently in ceramics of various types — these materials are far less subject to abrasion and wear than steel or brass. Cheap nozzles made of inferior materials soon abrade and then as the apertures enlarge the rate of spray application increases alarmingly. Fan nozzles are generally less subject to wear than cone nozzles, but cone nozzles usually give a better breakdown pattern of the water droplets.

The size of the apertures on the disc controls the rate of spray delivery, but below a certain minimum diameter the size does not greatly affect the size of the droplet produced. Some of the superior quality nozzles can be adjusted to obtain droplet sizes from a fine mist to a heavy drenching spray. Pesticides in suspension have to be used with particular care for nozzles with small apertures can easily be blocked by the particles. For many of the better makes of sprayers a range of nozzles is available so that the most suitable nozzle for a particular purpose can be employed.

Air-assisted or atomising machines (mist blowers) possess simple nozzles. The shearing force of the air produces the droplets. Spinning-disc ULV sprayers have no nozzles as the centrifugal force from the spinning disc produces the droplets.

The choice of spraying methods often depends upon the physical characteristics of the crop and the land on which it is grown. Hand machines are satisfactory for small areas and small crops. A high degree of efficiency is possible with hand-directed pesticide application. However, crops such as coffee, citrus, or rubber grown on a large scale demand large-scale application techniques such as tractor-drawn sprayers. In many regions, though, such machines cannot operate on the local terrain, because of steep slopes, rocks, etc. Aircraft spraying may solve the problem in these conditions, although again, fixed-wing aircraft are often of limited use in hilly terrain where flying conditions are tricky. They are more suited to extensive areas of monoculture. Often crops must be grown in such a way that they can easily be sprayed. This usually means adequate spacing and pruning, and may require some land preparation beforehand. Mechanised chemical pest and disease control on a large scale works only where the whole farming process is mechanised.

Hand-operated (manual) sprayers

Hand-operated spraying machines possess simple hydraulic nozzles and are usually of the knapsack design if they hold an appreciable volume of liquid. Some types can be pressurised by pumping up before spraying. There is a very wide range of models available suitable for a variety of purposes. Knapsack sprayers usually produce a coarse spray and it is necessary to apply a fairly large volume of liquid to ensure adequate coverage. Since they are operated by hand, greater accuracy in placing the pesticide on target can be achieved.

Compression systems These are operated by a compressed air pump.

1 *Atomisers* These consist of a simple compression cylinder with an inlet at one end for the air and an outlet at the other for the compressed air. Valves are seldom present. The outlet tube is fixed at right angles to a fine tube leading from the liquid container. On the compression stroke the air is forced across the open end of the feed tube and creates a vacuum which draws up the spray liquid from the tank. As the liquid is drawn up it is broken up into tiny droplets by the air stream. Hand atomisers are useful for treating individual plants, but they are tiring to operate for long periods. On the more refined atomisers the spray is delivered continuously by means of a pressure build-up system.

2 *Pneumatic hand sprayers* These are machines with a tank capacity varying from 0·5 to 3·5 litres where the tank acts as a pressure chamber. An air pump is attached to the chamber and it projects inside. The outlet pipe runs from the bottom of the tank to an external hydraulic nozzle. Air is pumped into the tank which compresses the liquid and forces it out of the nozzle when the release valve is opened to produce a continuous fine spray. The better machines can deliver a continuous spray for up to 5 minutes when fully charged with compressed air. These sprayers are most useful in glasshouses or for treatment of individual bushes under calm conditions. As with the previous type, these sprayers use very fine nozzles, they are thus more suitable for use with solutions or emulsions than suspensions, which may block the aperture.

Atomisers and pneumatic hand sprayers are not widely used by commercial growers.

3 *Knapsack pneumatic sprayers* (Fig 15.1 (b)). These appliances are basically the same as pneumatic hand sprayers except that they are designed for spraying large quantities of liquid (tank capacity up to 23 litres). The tank is usually carried on the operator's back, suspended on a harness with shoulder straps. The outlet pipe is extended by means of flexible tubing and terminates in some form of hand-lance. The lance usually carries from 1 to 4 nozzles and is easily carried in one hand. A hand valve on the lance base controls the flow of liquid. The air pump is operated with the sprayer on the ground. A high pressure is built up which will last for about 10 minutes of operation. These sprayers are manufactured in a wide variety of models, of varying degrees of efficiency. In general, they are very useful, especially for the small farmer or for trial work with pesticides. They can be very effective for estate work when teams of operators are employed, and individual attention to the plant is required. Since no system of agitation is incorporated, knapsack sprayers are more suitable for use with solutions than with suspended materials. Very long lances can be obtained for use in orchards and plantations.

Pump systems These are operated by a simple hydraulic pump.

1 *Syringes* Syringes consist of a cylinder into which the spray liquid is drawn on the return stroke of the plunger, and expelled on the compression stroke. The spray is sucked in through the spray nozzle aperture. In some cases there is a separate inlet near the nozzle controlled by a ball valve. The spray produced is drenching, and the syringe is difficult and tedious to use but they can be useful for spraying small numbers of plants. Most syringes are simple in construction, and will last for years with minimum maintenance.

2 *Force-pump sprayers* These are sprayers with a hand-operated pump. They have a lance and nozzle outlet and a feed-pipe to draw the spray liquid from a separate container. Although small in size these sprayers (fitted with a 46 cm double-action pump) can throw a jet of spray up to a height 12 m. These sprayers are good for spot treatments in orchards, and provided that the solution is kept stirred, they will spray suspensions as well as solutions and emulsions. This type of sprayer is obviously tiring to use. It is quite difficult to control the rate of application, but due to the double-action pump the spray is continuous.

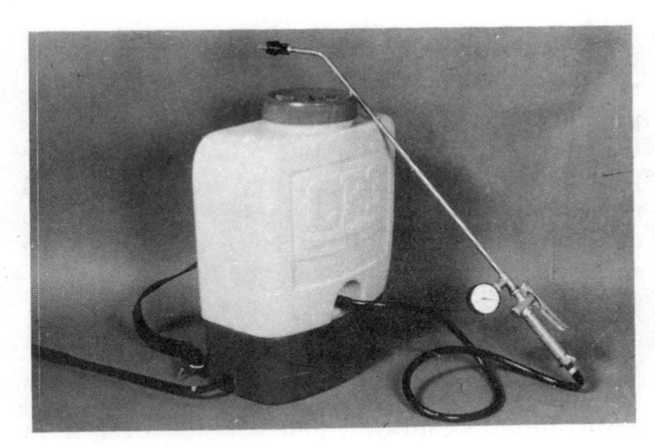

(a) A knapsack sprayer

fig. 15.1 (b) A knapsack compression sprayer used for spraying cotton in Thailand

3 *Stirrup-pump sprayers* These consist of a double-action pump suspended in a bucket. For support there is a foot stirrup reaching to the ground on the outside. A flexible outlet pipe carries the spray liquid from the pump to the spray lance which may vary in length and arrangement of nozzles. A stirrup-pump sprayer requires two operators, one to hold the lance and direct the spray, the other to stir the solution (if it is a suspension) and to work the pump. They are very useful, all-purpose machines, of robust construction which will withstand hard wear. Bush crops, buildings and small trees can be sprayed easily with a stirrup-pump sprayer, and it is ideal for team operation. Providing the liquid is kept stirred, quite coarse suspensions can be sprayed. A large version, mounted on wheels, with a large-capacity double-action pump is available for treating larger areas.

4 *Knapsack sprayers* (Fig. 15.1(a)) These all-purpose, successful, sprayers are used throughout the world for spraying pesticides over smaller areas. They consist basically of a spray container which sits comfortably on the back of the operator, held by shoulder straps. The double-action pump is built either inside or outside the spray container. It is operated by working a lever which projects alongside the operator's body. In some models the pump lever also operates an agitation paddle in the spray tank. The spray liquid is applied through a lance held in the operator's free hand; the lance is connected to the spray tank by a long flexible hose. The tank capacity is usually about 20 litres. Provided that a sufficiently coarse spray nozzle is used this sprayer can be used with any type of spray.

Many of the most recent sprayers are almost entirely made of plastic which obviates the problem of metal corrosion by the more corrosive pesticides. With a little practice the rate of spray application can be controlled quite accurately. Knapsack sprayers can be tiring to operate over a long period of time, but they are very versatile, quite robust in construction and portable. They are useful on small farms or in teams on larger estates, or for pesticide trials.

Various modifications can be made to the nozzles and lance arrangements of these machines (Fig. 15.2). For spraying of cotton in Malawi, a system of laterally projecting nozzles on a vertical lance is used. The lance is attached to the machine on the operator's back.

Power-operated sprayers I Hydraulic mechanisms

Compression systems

1 *Hand guns* Two types of compression hand-spray guns are made. One type has the spray liquid fed into a pipe through which air from a portable compressor is fed. The other type is similar to the small compressed air sprayer that is worked by hand, but the air pump is replaced by an inlet from a portable compressor. Both types can only operate at the length of the lead from the compressor. Droplet size and rate of application can usually be carefully controlled, but the capacity of the spray tank is small and hand guns are only of value where small areas have to be covered with small amounts of spray. Suspensions may block the outlet nozzle especially if it is of very small aperture. The advantage of these sprayers is that any type of compressor can be used, and they can also be used as paint sprayers.

2 *Portable sprayers* Many types of small portable sprayers are manufactured, some can be carried easily by one man and others are larger and mounted on a wheeled chassis. Air is compressed by a small compressor and is forced into the spray container. The container usually holds about 45

fig. 15.2 A knapsack hydraulic sprayer with nozzles mounted on a frame carried on the operator's back

litres, and is of strong, welded construction. It is operated at a pressure of about 7 kg per square centimetre. The outlet hose from the spray container may end either in single or multiple lances, or in a boom. Provided the spray tank is lined with an anti-corrosive material, these sprayers can be used for spraying corrosive liquids, since there is no pump to be corroded. As the air compressor can be used for other purposes these machines can be useful on small farms where versatility is required.

3 *Large mounted sprayers* These are basically similar to the smaller portable sprayers, but they have spray tanks of a much larger capacity, hence they require larger compressors. The whole machine is usually pulled by a tractor, and the outlet terminates in a spray boom of varying design. Sometimes the outlet is a series of hand lances which can be used to spray trees. (Fig. 15.3). The spray booms may cover 6–9 m, or more, in a single swath. The better booms usually can be adjusted for height to suit the crop being sprayed. Rates of application can be adjusted by altering the size of the jet aperture in the nozzles. The booms may be positioned vertically to spray fruit crops.

fig. 15.3 Spraying citrus in Central America with hand lance operated from a motorised hydraulic sprayer

Pump systems

1 *Portable sprayers* The range of small portable sprayers available is now very extensive. They are all based upon the fundamental units of a power source, a pump and a spray tank. The smallest units consist of a small double-action reciprocal pump harnessed to a small air-cooled engine. This is mounted on a framework so that it can be conveniently carried. The largest units may include a tank complete with positive agitator mounted upon a four-wheeled chassis. The pump outlet pipe may supply a number of hand lances, depending upon the capacity of the pump. The mounted machines may have a small boom mounted so that they can spray ground crops. These sprayers are only really suitable for small areas of orchard or plantation crops, as they are generally limited effectively to two lances.

2 *Large sprayers* Tractor-driven hydraulic machines are used usually for high-volume spraying. They apply the diluted pesticide through a set pattern of spray nozzles attached to a boom. The spray pattern may be applied over the top of the crop as a dense mist, or projected laterally and vertically into the foliage of tree crops. They have a range of tank capacities of 180–1 800 litres. They can operate at very high pressures (56 kg cm^{-2}) with a high rate of delivery. They are invariably mounted on tractors and often have booms, which will deliver from 57 to 3 370 litres per hectare. Some of the larger booms are built vertically so that the nozzles point upwards. This ensures that the under-leaf is sprayed as well as the top surfaces; this is particularly useful for crops such as coffee. These sprayers may be mounted on aircraft for low-volume aerial use (see Fig. 14.4 on p. 80).

Power-operated sprayers II Atomisers

Air-assisted machines or atomisers have different nozzles from hydraulic machines. The spray is made by the shearing force of a high-velocity air stream over a simple tubular nozzle; this produces finer droplets. These are efficiently transported to the target by the large volume of air which the machine forces out by powerful fans. Motorised knapsack mist blowers or atomisers are very popular hand-directed types of spraying machine and give very good spray cover when properly used. There are many different makes of tractor-drawn air-blast machine, many of which are used on estate-grown perennial crops such as coffee and citrus.

Rotary atomisers produce very fine droplets within a narrow size range. These are used chiefly in ULV spraying from aircraft, but this can also be done by using fine hydraulic nozzles. Because of the cost of carrying large volumes of liquid by air, ULV spraying was developed specifically for aircraft use, so that small volumes of pesticide, usually in concentrated form (i.e. with little carrier) can be applied over a large area. Fixed-wing aircraft are usually used with the spray nozzles mounted under the wings. Helicopters have also been used but are more expensive to operate.

Mist blowers Low-volume mist blowers and fogging machines are of relatively recent development.

The air-blast systems consist basically of a series of nozzles which produce a coarse spray through the wide apertures. This is then broken into fine droplets by a fast-moving air stream. This basic system has many modifications, mainly in the method in which the liquid is introduced into the air stream to be broken up. The air stream is usually produced by a centrifugal fan of large capacity. Mist blowers vary enormously in tank size from portable machines that can be carried by one man to large self-mounted tractor-drawn machines.

1 *Motorised knapsack mist blowers* (Fig. 15.4) These machines consist of a small motor unit, housing and spray tank, mounted on a knapsack. A wide, flexible air-tube connects with the nozzle end to produce the droplets. These are popular and efficient machines although more expensive than hydraulic knapsack sprayers. A number of different types are made by different manufacturers. The types differ in efficiency, especially when

fig. 15.4 Motorised knapsack mist blower

fig. 15.5 Tractor-drawn mist blower

spraying upwards into tree crops. Some machines have nozzle arrangements which enable them to use very low volumes, down to about 15 litres per hectare.

2 *Tractor-powered mist blowers* (Fig. 15.5) These are very popular for spraying plantation crops. They have replaced some of the older hydraulic machines. Some have a very large capacity, they project a mist all around the housing. Others have adjustable air tubes which allow more control over spray direction. Spray drift is the main hazard arising from the use of these machines.

Ultra-low-volume sprayers These machines produce very fine droplets ($<70\ \mu m$ diameter) within a narrow size range. They are now used manually and in tractor-mounted equipment as well as aircraft.

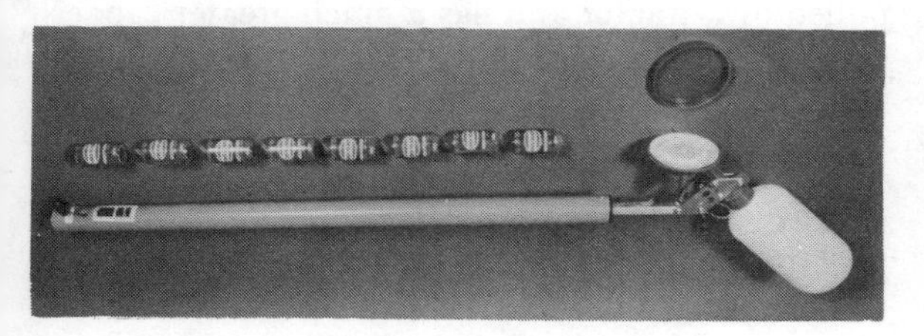

(a) A micron ULVA dismantled

fig. 15.6 ULVA sprayers
(b) A ULVA sprayer in use for cotton in Malawi

The droplets are carried in a light oil; they are either blown by a fan on to the crop or allowed to settle naturally in drift spraying. The basic unit for producing the spray is a rotary atomiser. This consists of a spinning disc or similar structure, often with a serrated edge, into which the spray liquid is fed. The droplets are produced as the liquid is projected from the edge of the spinning disc by the centrifugal force generated by its rotation.

Small hand units operated by battery are available, e.g. ULVA (Figs 15.6 (a) and (b)). In some models a motor-driven fan mounted behind the disc propels the droplets. Much ULV spraying is done from aircraft and the apparatus used consists of spinning 'cages' mounted beneath the wings as in the 'Micronair' unit (Fig. 15.7). Tractor-mounted ULV sprayers (e.g. Mantis) may have a series of spinning discs mounted in front of a fan which blows the fine drops into the crop canopy.

Controlled droplet application (c.d.a.) The range of sizes and density of droplets in a spray are vital factors in determining the effectiveness of a particular treatment. Fields of crops may be regarded as flat surfaces, in fact dosage should be adjusted to allow for the difference between area of foliage and ground area. Spray deposited by conventional equipment, for both ground and aerial application, falls mainly within the general target area, i.e. the field where the crop is growing, but some droplets drift outside the target area. Drift within the crop canopy leads to a very low efficiency of spraying (less than 0·1 per cent). The most effi-

fig. 15.7 Micronair unit mounted on an aircraft

cient spraying recorded was 6 per cent on aerial locust swarms.

Droplet size and density can be controlled to give effective penetration of the crop foliage even with minimal volumes. Rotary atomisers allow more precise control of droplet size and density than the other spraying mechanisms. This minimises wasteful spray drift while ensuring maximum deposition on the target.

Dusters

All dusters consist basically of a hopper (a container for the dust), a system of agitation to disturb the dust and a feed mechanism to pass the dust into the current of air which is carried through an outlet as a turbulent cloud.

Hand dusters are usually primitive in structure and tedious to use, but can be effective for the small farmer. They may consist of a crude distributing arrangement built as part of the packaging. Either two concentric cylinders free to move within each other, or a cardboard piston or diaphragm for pumping the air. As the air passes through the pack it picks up a small quantity of dust and ejects it through a nozzle.

Hand-pump dusters are cheap and easy to operate but cannot control the amount of dust delivered. Most advanced pumps have a double-action plunger which maintains a constant and even stream of air.

Bellows-type dusters generate the air-stream by the contraction and expansion of a pair of bellows. The commonest types are worn on the back in knapsack fashion. The dust hopper, containing from 4-7 kg of dust, is carried on the back in a metal cradle. The bellows are situated either on the back or top of the hopper and drive a stream of air through a tube into a small mixing chamber by the hopper outlet. The dust is fed into the mixing chamber by a simple agitator. The air and dust travel along a flexible pipe running along the side of the operator. This can be controlled easily with one hand. The outlet may vary in shape or design for different purposes. The bellows and agitator are operated by a simple up and down movement of a lever worked by the free hand.

Rotary hand dusters produce an air stream by a fan driven off a hand crank through a reduction gear. They are often mounted on the chest or back of the operator, using a metal frame and system of straps.

Power dusters are manufactured in a variety of forms. Traction dusters derive their motive power from the turning of the land wheels — either as a wheelbarrow type or a two-wheeled trailer type. Wheelbarrow-type dusters are suitable for use on small areas only, whereas the trailer type is usually pulled by a tractor and has a much greater capacity. Power dusters are equipped with independent engines to provide the power for their operation. The smallest types can be strapped to the chest of the operator, but the larger types become very expensive. The larger power boom dusters are effective for treatment of large areas. Some of these dusters work on the drift principle and use the movement of ambient air to distribute the powder over the crop. Aeroplane dusting is carried out in areas where the terrain and crops are suitable and again the air stream is used to spread the powder over the crop.

Thermal fogging Some pesticides can be vaporised in an oil suspension by heat to produce a fog or smoke. Hot exhaust gases from engines may be used to produce the fog which is then allowed to drift through foliage. This is a useful method for fumigating dense scrub or crops. Portable thermal foggers are also made and have been used to control rubber tree diseases and a wide variety of pests.

16 Efficient and safe use of pesticides

Protection of the environment

The control of plant diseases and pests by the use of chemicals must be carried out with great care. The chemicals commonly used for plant protection are reasonably safe provided that the precautions which manufacturers list on the labels are followed. There is now considerable pressure to use chemicals which are not dangerous to man, wildlife or any other aspect of the environment. Many countries have legislation which only permits chemicals approved for certain uses to be freely available on the market and licences may be required to use certain toxic chemicals. Other countries, e.g. the UK, have a voluntary approval scheme in which manufacturers are invited to have their products approved by an impartial body of government scientists. Approved chemicals will always have a wider market so that the scheme is in the interest of pesticide manufacturers.

Conflict arises when choices of the following kinds are available: a well-established, cheap and efficient pesticide, which, if used carelessly, may constitute an environmental risk; and a new, more expensive pesticide which is environmentally safe, but which is less efficient or of more limited use. This conflict leads to difficulty in giving approval to a pesticide. It is particularly acute in developing countries where resources are limited and the need for greater agricultural production is desperate.

Application

With many of the newer, short-lived organophosphorus and carbamate chemicals dosage, placement and timing of application are critical. Accurate rates of application are essential. Thus, attention should be paid to maintenance of equipment for defects such as faulty or worn nozzles which can appreciably alter rates of application.

Compatibility Compatibility of mixed products used in combined spray programmes is important. If no information is given on the labels consult the manufacturer or agent. Wetting agents are sometimes used as additives to sprays for crops that are difficult to wet (e.g. brassicas, peas), and not all wetting agents are compatible with all proprietary formulations.

Damage to plants Taint and damage to plants are hazards to be avoided. Where chemicals are known to cause damage to certain crops or varieties of plants, or if taint or off-flavours are produced in edible crops, this is mentioned in Chapters 17 and 18 in details of the specific chemicals.

Pest resistance Pest resistance to chemicals is one of the major problems in agricultural entomology in Europe and North America, and has recently become evident in parts of Africa and several Asian countries. Such cases arise initially in localised areas or even on certain holdings and may remain purely local for some time. Resistance does not develop until a particular chemical has been employed in that area for 3 - 10 years. Sometimes resistance to one compound may be closely followed by resistance to other related compounds, e.g. one organochloride compound and then others. In some areas of the Tropics pest resistance is now of great importance due to the almost continuous breeding of some pest species.

Resistance has also developed in fungi to some of the systemic fungicides, notably the MBC type fungicides, where these have been used intensively.

Storage Storage of chemicals is important; they should be kept in a dry place protected from extremes of temperature. Generally, frost is more harmful to liquid formulations than is high temperature. Most formulations should be usable for at least 2 years without loss of efficiency. However, unsatisfactory storage conditions can impair stability and effectiveness.

Correct use of pesticides

All pesticides should be treated with care whether they are known to be poisonous or not. For safe and efficient use a number of points need to be observed.

1 Always use the correct pesticide for the job. If in doubt consult plant protection experts.
2 Read the label carefully, especially the safety precautions, before use.
3 Always use the correct dosage. The amount of pesticide to be used for a particular purpose on a specific crop can be expressed in different ways. As weight (kg) per unit area (ha), or as weight (kg) mixed with a specific volume (litres) of water for use per unit area (ha). Manufacturers instructions or expert advice should always be followed. Too small a dose may result in poor control, whereas too high a dose may damage the crop.

4 Do not use persistent chemicals when there are effective, less persistent alternatives.
5 Always use the correct method of application. Pesticides are designed to be applied in specific ways, e.g. to soil, seed, foliage, etc., with specific apparatus. The way in which they are applied affects their efficiency and safety.
6 Correct timing of application, in relation to the stage of crop growth, appearance of pest or pathogen, and the frequency of application is vital for efficient control.
7 Avoid drift on to other crops, livestock and neighbouring property and take care to prevent contamination of any water source, whether used for drinking or irrigation purposes.
8 Safely dispose of all used containers. Liquid contents must first be washed out thoroughly and the washing added to the spray tank. Packages containing powders or granules must be completely empty before disposal. Burn bags, packets and polythene containers. Puncture non-returnable metal containers (except aerosol dispensers) and bury them in a safe place. Bury glass containers or dispose of them with other refuse. On no account use empty pesticide containers for any other purposes.
9 Return unused materials to a chemical store and keep under lock and key.
10 Clean any protective clothing used, and wash exposed parts of the body thoroughly when the job is completed.

Characteristics of the chemical Particular attention should be paid to the paragraphs headed 'Caution' under the individual pesticides listed in Chapters 17 and 18. It gives the following indications.

1 Whether the chemical is particularly poisonous, if so, then certain protective clothing should be worn.
2 Whether there are any special risks to the user, even if not very toxic, such as irritation to the skin. Some protective clothing should be used with these chemicals, and their labels should be consulted for guidance.
3 What precautions should be taken to ensure that unacceptable residues do not remain on edible crops at harvest. Where appropriate, the time at which a chemical may be applied is shown and also the minimum interval that must be observed between last treatment and harvest.
4 Whether there are risks to bees, fish, livestock, or other wildlife. In the case of bees the degree of risk is given as dangerous, harmful and safe (see below). Fish are particularly susceptible to many pesticide chemicals, and great care must be taken to prevent contamination of ponds, waterways and ditches with chemicals or used containers.

Toxicity of pesticides to bees

Dangerous Highly toxic to bees working the crop or weeds, at the time of treatment, and toxic for 24 hours or more.

Azinphos-methyl	Fenitrothion
BHC	Formothion
Carbaryl	Heptenophos
Chlorpyrifos	Lead (II) arsenate (V)
Demeton-S-methyl	Methomyl
Sulphone	Mevinphos
Diazinon	Parathion
Dichlorvos	Permethrin
Dimethoate	Phosphamidon
Endrin	Pirimiphos-methyl
Ethoate-methyl	Pyrazophos
	Triazophos

Harmful Toxic to bees working a crop at time of treatment, but not hazardous if applied when bees are not foraging.

Benodanil	Mercury compounds
DDT	Nicotine
DDT with malathion	Omethoate
Demephion	Oxydemeton-methyl
Demeton-S-methyl	Phosalone
Endosulfan	Propoxur
Fenitrothion	Tetrasul
Malathion	TDE
Methidathion	Thiometon
Menazon	Vamidothion

Safe

Binapacryl	Disulphoton granules
Derris	Phorate granules
Dicofol	Schradan
Dinocap	Tetradifon

The Agriculture (Poisonous Substances) Regulations for the UK

These regulations are laid down in the Act of 1952 and although not applicable outside the UK they give a good indication of the toxicity of the various chemicals involved and the protective clothing ideally required. It must be noted, though, that this clothing is designed for use in a temperate climate and not in the Tropics, especially the more humid regions.

Part I Substances

Dimefox and chloropicrin are no longer approved for use in the UK.

Full protective clothing, e.g. rubber gloves and boots, respirator and with an overall and rubber apron, or a mackintosh when preparing the diluted chemical. The respirator may be dispensed with when the diluted chemical is applied to the soil.

Part II Substances

Aldicarb	Methomyl
Carbofuran	Mevinphos
Disulfoton	Oxamyl
DNOC	Parathion
Endosulfan	Phorate
Endrin	Schradan
Fonofos	Thionazin

Full protective clothing, which includes rubber gloves, boots, either a face shield or dust mask, and an overall and rubber apron or a mackintosh; or a hood, or a rubber coat and sou'wester, depending on the operation being performed. A respirator is required when applying aerosols or atomising fluids in glasshouses.

When the substances are used in the form of granule, certain relaxations in the protective clothing requirements are now allowed.

Part III Substances

Azinphos-methyl	Methidathion
Chlorfenvinphos	Nicotine
Demephion	Oxydemeton-methyl
Demeton-*S*-methyl	Phosphamidon
Drazoxalin	Thiometon
Fentin	Vamidothion
Dichlorvos	

Rubber gloves and face shield when preparing the diluted chemical. Full protective clothing, i.e. overall, hood, rubber gloves and respirator, is needed when applying aerosols or atomising fluids in glasshouses.

The way in which a pesticide is applied often influences how hazardous it is to the operator or to the environment. Granular formulations usually present least hazards especially if they are non-dusty. If handled with care (rubber gloves should always be used) there is little danger of them contacting sensitive parts of the body. At the other extreme, application methods which produce very small droplets of particles such as fogging, dusting, or ultra-low-volume spraying can be very hazardous even with mildly toxic chemicals. This is because the fine particles or droplets in aerial suspension consist of highly concentrated chemicals and are easily inhaled deep into the respiratory system. Only formulations recommended for these application methods should be used and every precaution should be taken when using them — especially to prevent inhalation of fine particles.

Pesticides/pest charts

Only major pests have specific control recommendations. If other insects must be killed it can be borne in mind that some pesticides do tend to be toxic to groups of related insects. If no recommendation is available then Tables 16.1, 16.2 and 16.3 may be of use in deciding which chemical to use. This is not always successful, for although some chemicals are specific to certain groups of insects, there are notable exceptions. For example, carbaryl is a pesticide generally effective against caterpillars in most parts of the world. In Malawi on the cotton crop it is effective against red bollworm (*Diparopsis castanea*) but has no appreciable toxic effect upon *Heliothis* spp. caterpillars. DDT has to be used against the latter pest. A spray mixture is used against a mixed infestation.

Table 16.1 Pests

Chemical	Locusts	Termites	Homoptera	Aphids	Mealybugs	Scales	Heteroptera	Thrips	Caterpillars	Fly larvae	Beetles	Mites	Miscellaneous
Chlorinated hydrocarbons													
Aldrin	•	•	•		•				•	•	•		
BHC			•	•			•	•		•	•		Ants
DDT	•	•	•	•	•	•	•	•	•	•	•		
Dicofol												•	
Dieldrin	•	•	•	•	•	•	•	•	•	•	•		Ants
Endosulfan			•	•			•				•	•	
Endrin	•	•	•	•	•	•	•	•	•	•	•	•	
Heptachlor										•	•		
Mirex													Ants
Tetradifon												•	
Tetrasul												•	
TDE							•	•	•		•		
Substituted phenols													
Amitraz												•	
Binapacryl												•	
DNOC	•		•	•	•	•	•		•			•	
Pentachlorophenol		•									•		
Organophosphorus													
Azinphos-methyl	•		•	•	•	•	•	•	•	•	•	•	
Azinphos-methyl with demeton-S-methyl sulphone			•	•				•	•	•	•	•	
Bromophos			•						•	•	•		
Bromopropylate				•	•	•				•		•	
Chlorfenvinphos				•								•	
Carbophenothion			•	•	•	•	•					•	
Chlorpyrifos			•	•	•	•	•					•	
Demephion			•	•			•	•		•		•	Ants
Demeton	•		•	•	•	•	•	•	•	•	•	•	Ants
Demeton-S-methyl			•	•								•	
Diazinon			•	•				•		•		•	
Dichlorvos			•	•						•	•	•	
Dimefox			•	•		•				•		•	
Dimethoate				•						•		•	
Disulfoton	•		•	•	•	•	•	•	•	•	•		
Ethion			•	•			•	•	•	•	•		
Ethoate-methyl				•						•		•	
Fenitrothion	•		•	•	•	•	•	•	•	•	•		
Fenthion			•	•			•	•	•	•	•		

Table 16.1 Pests (continue)

Chemical	Locusts	Termites	Homoptera	Aphids	Mealybugs	Scales	Heteroptera	Thrips	Caterpillars	Fly larvae	Beetles	Mites	Miscellaneous
Fonofos									•	•	•		Birds
Formothion			•	•								•	
Heptenophos			•	•	•	•	•	•	•	•	•	•	
Malathion			•	•	•	•	•			•		•	
Mecarbam				•									
Menazon			•	•	•	•	•		•			•	
Methidathion			•	•					•	•	•	•	
Mevinphos			•				•		•		•	•	
Monocrotophos			•							•		•	
Naled			•							•		•	
Omethoate	•		•	•	•	•		•	•		•	•	
Oxydemeton-methyl			•	•				•		•		•	
Oxydisulfoton			•	•								•	
Parathion			•	•	•	•	•			•		•	
Parathion-methyl			•	•	•	•	•			•			
Phenthoate			•	•					•		•	•	
Phorate			•	•			•			•	•		
Phosalone			•	•					•		•	•	
Phosmet			•	•			•			•	•		
Phosphamidon			•	•				•	•	•	•	•	
Phoxim												•	
Pirimiphos-ethyl	•		•	•		•	•	•	•	•	•	•	
Pirimiphos-methyl	•								•	•	•		
Profenofos			•						•	•	•	•	
Prothoate	•		•	•	•	•		•	•	•	•	•	Ants
Quinomethionate			•				•	•				•	
Schradan												•	
TEPP				•								•	
Terbufos				•								•	
Tetrachlorvinphos									•	•			
Thiometon				•							•		
Thionazin										•		•	
Thioquinox												•	Nematodes
Triazophos				•				•		•		•	
Trichloronate										•	•		
Trichlorphon			•				•		•	•	•		
Tricyclohexyltin hydroxide												•	
Vamidothion			•	•								•	

Table 16.1 Pests (continue)

Chemical	Locusts	Termites	Homoptera	Aphids	Mealybugs	Scales	Heteroptera	Thrips	Caterpillars	Fly larvae	Beetles	Mites	Miscellaneous
Carbamates													
Bufencarb			●			●	●		●		●		Sawfly larvae
Carbaryl	●		●			●	●		●	●	●		
Carbofuran			●	●					●	●	●	●	
Methiocarb				●					●		●	●	Slug
Methomyl									●		●	●	
Oxamyl				●				●		●	●	●	Nematod
Pirimicarb				●						●			
Promecarb									●	●	●		
Propoxur		●	●	●			●			●		●	Ants
Miscellaneous compounds													
Aluminium phosphide									●		●		
Bromomethane									●		●		Nematod
Dibromoethene											●		
Copper(II) acetoarsenite									●		●		
Lead arsenate(V)									●	●			Sawfly larvae
Mercury(I) chloride										●			
Sulphur (lime)												●	
Natural organic compounds													
Bioallethrin										●			
Nicotine				●			●	●		●			
Pyrethrins			●				●		●	●		●	Sawfly larvae
Rotenone				●				●	●		●	●	
Organic oils													
Petroleum oils					●	●		●				●	
Tar oils			●	●	●	●	●		●				Eggs
Biological compounds													
Bacillus thuringiensis									●				
Diflubenzuron			●						●			●	
Methoprene										●			

Addendum

Cypermethrin ('Cymbush'; 'Ripcord')
A synthetic pyrethroid, broad-spectrum, with both contact and stomach action effective against leaf and fruit-eating Lepidoptera and Coleoptera, as well as Hemiptera, and animal ectoparasites; widely used against cotton pests, and on many tropical and temperate crops.

Table 16.2 Crop Guide For Pesticide Use

Crop	Major pest	Effective chemicals
Avocado	Citrus blackfly	Dimethoate
	Fire ant	Dieldrin, BHC, diazinon
Bananas	Banana aphid	Parathion, phosphamidon, endrin, dicrotophos
	Banana fruit-scarring beetle	Aldrin, dieldrin, endrin
	Banana scab moth.	DDT, DDT/BHC
	Banana thrips	DDT, dieldrin, BHC, phosphamidon
	Banana weevil	Aldrin, dieldrin
Brassicas	Cabbage aphid	Dimethoate, demephion, demeton-S-methyl, dichlorvos, disulfoton, formothion, malathion, menazon, schradan, etc.
	Cabbage sawfly	DDT, carbaryl, pyrethrum
	Bagrada bugs	DDT, BHC, carbaryl
	Diamond-back moth	DDT, carbaryl, malathion, pyrethrum
Brinjal	Blister beetle	Dieldrin, parathion
	Epilachna beetle	Dieldrin, malathion
	Potato tuber moth	DDT, dicrotophos, dimethoate, parathion
Capsicums	Cotton aphid	DDT, DDT/carbaryl, carbaryl, dimethoate
	Peach-potato aphid	Dimethoate, malathion, demeton-S-methyl, menazon, demephion, disulfoton, formothion, phorate, phosphamidon, thiometon
	Red cotton mite	Dimethoate, dicofol, tetradifion, etc.
	Striped blister beetle	Dieldrin, parathion
	Yellow tea mite	Dicofol
Cashew	Cashew helopeltis	BHC
	Cashew stem girdler	(cultural control only)
	Cashew weevil	(cultural control only)
	Citrus blackfly	Dimethoate
Cassava	Cassava scale	(not usually required)
	Elegant grasshopper	Aldrin, BHC, dieldrin, carbaryl
	Tobacco white-fly	DDT, dimethoate, pyrethrum

Castor	Castor stem borer	(not usually required)
	Cotton helopeltis	DDT
	False codling moth	DDT, dichlorvos, mevinphos, parathion, phosphamidon
	Green stink bug	DDT, BHC, phorate
Citrus	Black tea thrips	Fenitrothion
	California red scale	Malathion, white oil, diazinon.
	Citrus aphids	Dimethoate
	Citrus blackfly	Dimethoate
	Citrus bud mite	Lime-sulphur, chlorobenzilate
	Citrus (root) mealybug	Aldrin
	Citrus psyllid	Dimethoate
	Citrus red spider mite	Dimethoate, dicofol, tetradifon, etc.
	Citrus rust mite	Lime-sulphur, chlorobenzilate
	Citrus thrips	Lime-sulphur
	Citrus whitefly	Azinphos-methyl, trichlorphon, parathion-methyl
	Cottony cushion scale	Azinphos-methyl, parathion-methyl
	False codling moth	DDT, dichlorvos, mevinphos, parathion, phosphamidon
	Long-tailed mealybug	(not usually required)
	Mediterranean fruit fly	Malathion, trichlorphon, fenthion
	Mussel scale	Diazinon, malathion, carbaryl
	Olive (black) scale	Azinphos-methyl, diazinon, malathion, parathion, parathion-methyl, white oil
	Orange dog	Fenitrothion, fenthion, malathion
	Purple scale	Diazinon, malathion, carbaryl, parathion
	Red Crevice Mite	Dicofol, chlorobenzilate
	Soft green scales	Diazinon, malathion, dimethoate (dieldrin against ants)
Cocoa	Black tea thrips	Fenitrothion
	Citrus aphid	Dimethoate
	Citrus (root) mealybug	Aldrin
	Cocoa capsid	BHC
	Cocoa stem borer	DDT, dieldrin, endrin, phosphamidon
	Cotton helopeltis	DDT
	Soft green scale	Diazinon, malathion, dimethoate
Coconut	Black line scale	Diazinon, ethion, parathion, white oil
	Coconut bug	Dieldrin (to kill the ants)
	Coconut scale	Malathion, diazinon, parathion, dieldrin
	Coconut weevils	Dieldrin, tars
	Palm weevils	Aldrin, dieldrin, demeton-S-methyl
	Rhinoceros bettles	(cultural control preferred)

Coffee	Antestia bugs	Fenitrothion, fenthion
	Black borers	Dieldrin
	Citrus (root) mealybug	Aldrin
	Coffee berry borer	Dieldrin, BHC
	Coffee berry moth	Fenithrothion, fenthion
	Coffee leaf miners	Fenithrothion, fenthion
	Coffee leaf skeletoniser	Dieldrin, trichlorphon
	Coffee leaf-rolling thrips	(not usually required)
	Coffee lygus (capsid)	Fenitrothion, fenthion, pyrethrum
	Coffee thrips	DDT, dieldrin
	Dusty brown bettle	Aldrin, dieldrin
	Helmet scale	White oil, (dieldrin — against ants)
	Jacaranda bug	(not usually required)
	Kenya mealybug	Dieldrin (to kill the ants)
	Lace bugs	Fenitrothion
	Red coffee mite	Dimethoate, dicofol, tetradifon, etc.
	Soft green scales	Diazinon, malathion, dimethoate, (dieldrin — against ants)
	Star scale	Tar oil
	White waxy scale	White oil, carbaryl, dimethoate, malathion, azinphos-methyl, carbofenthion
	Stinging caterpillar	Fenitrothion, fenthion,
	Striped mealybug	Malathion, azinphos-methyl
	Systates weevil	DDT, dieldrin
	White coffee borer	Dieldrin
	Yellow-headed borer	Dieldrin
Cotton	American bollworms	DDT, BHC, DDT/BHC
	Spiny bollworms	Carbaryl, DDT/BHC
	Blue bugs	DDT
	Cotton aphid	DDT, DDT/carbaryl, carbaryl, dimethoate
	Cotton helopeltis	DDT
	Cotton jassids	DDT, BHC
	Cotton leaf roller	DDT, carbaryl
	Cotton lygus	DDT
	Cotton semi-looper	Diazinon, parathion, carbaryl
	Cotton seed bug	BHC
	Cotton stainers	Carbaryl, BHC/DDT
	False codling moth	DDT, dichlorvos, mevinphos, parathion, phosphamidon
	Green stink bug	DDT, BHC, phorate
	Pink bollworm	DDT/endrin, carbaryl, trichlorphon, azinphos-methyl

	Red bollworms	Carbaryl, endrin, DDT/toxaphene, monocrotophos
	Red cotton mite	Dimethoate, dicofol, tetradifon, etc.
	Tobacco whitefly	DDT, dimethoate, pyrethrum
	Yellow cotton mite	Dicofol
Cucurbits	Epilachna	Dieldrin, malathion
	Melon fly	Malathion, trichlorphon, fenthion
	Leaf-footed plant bug	BHC, parathion
Groundnut	American bollworm	DDT, BHC, DDT/BHC
	Cotton leafworm	DDT, endosulfan, malathion, carbaryl, trichlorphon
	Groundnut aphid	Menazon, dimethoate
	Groundnut hopper	Dieldrin
	Lesser armyworm	DDT, BHC, endrin, parathion
	Striped sweet potato weevil	(not usually required)
	Systates weevil	DDT, dieldrin
Guava	Helmet scale	White oil, (dieldrin — against ants)
	Red-banded thrips	Fenitrothion
Mango	Mango fruit fly	Malathion, trichlorphon, fenthion
	Mango seed weevil	(not usually required)
	Red-banded thrips	Fenitrothion
Maize	African armyworm	DDT, endosulfan, malathion, carbaryl, trichlorphon
	American bollworms	DDT, BHC, DDT/BHC
	Bean seed fly	Dieldrin, ethion, diazinon, trichloronate
	Black maize beetle	Aldrin, dieldrin, DDT, BHC, chlordane
	Coastal stalk borer	DDT
	Epilachna beetle	Dieldrin, malathion
	European corn borer	Trichlorphon, parathion, parathion-methyl
	Maize aphid	Dimethoate, demephion, demeton-S-methyl, formothion, menazon, ethoate-methyl
	False codling moth	DDT, dichlorvos, mevinphos, parathion, phosphamidon
	Maize leaf-hopper	(not usually required)
	Maize stalk borer	DDT
	Maize tassel beetle	DDT
	Maize webworm	DDT, carbaryl, fenitrothion
	Maize weevil	Malathion, BHC, DDT/BHC
	Pink stalk borer	Endrin, dieldrin, parathion, diazinon, trichlorphon, dichlorvos

Crop	Pest	Insecticide
	Rice armyworm	DDT, BHC, endrin, parathion
	Rice stalk borer	Endrin, dieldrin, parathion, diazinon, trichlorphon, dichlorvos
	Shiny cereal weevil	DDT
	Sorghum shoot fly	Endosulfan, phorate, disulfoton
	Spotted stalk borer	DDT, endrin
	Sugar cane stalk borer	Endrin
Millets	Coastal stalk borer	DDT
	Cotton lygus	DDT
	Elegant grasshoppers	Dieldrin, aldrin, BHC, carbaryl
	Epilachna beetle	Dieldrin, malathion
	Pink stalk borer	Endrin, dieldrin, parathion, diazinon, trichlorphon, dichlorvos
	Shiny cereal weevil	DDT
	Sorghum shoot fly	Endosulfan, phorate, disulfoton
	Spotted stalk borer	DDT, eldrin
	Striped blister beetle	Dieldrin, parathion
Oil Palm	Coconut weevil	Dieldrin, tars
	Long-tailed mealybug	(not usually required)
	Palm weevils	Aldrin, dieldrin, demeton-S-methyl
	Rhinoceros beetle	(cultural control usually)
Okra	American bollworm	DDT, BHC, DDT/BHC
	Cotton jassid	DDT, BHC
	Cotton seed bug	BHC
	Cotton stainer	Carbaryl, DDT/BHC
	Spiny bollworms	Carbaryl, DDT/BHC
Onions	Bean seed fly	Dieldrin, ethion, diazinon, trichloronate
	Onion thrips	Dieldrin, diazinon, DDT, malathion
Papaya	Citrus whitefly	Azinphos-methyl, trichlorphon, parathion-methyl
Passion Fruits	Kenya mealybug	Dieldrin, (to kill the ants)
	Mediterranean fruit fly	Malathion, trichlorphon, fenthion
	Red crevice mite	Dicofol, chlorobenzilate
Pineapple	Pineapple mealybug	Malathion, diazinon, parathion
Potato	Colorado beetle	DDT, carbaryl, azinphos-methyl, phosphamidon
	Epilachna beetles	Dieldrin, malathion
	Peach-potato aphid	Dimethoate, malathion, demeton-S-methyl, menazon, demephion,

Crop	Pest	Insecticide
		disulfoton, phorate, formothion, thiometon,
	Potato aphid	Dimethoate, formothion, menazon, demeton-S-methyl, diazinon, pyrethrum
	Potato tuber moth	DDT, dicrotophos, dimethoate, parathion
Pulses	American bollworm	DDT, BHC, DDT/BHC
	Bean bruchid	BHC, pyrethrum
	Bean flower thrips	DDT/BHC
	Bean fly	Aldrin, dieldrin
	Bean seed fly	Dieldrin, ethion, diazinon, trichloronate
	Black bean aphid	Disulfoton, phorate, menazon, endosulfan, demeton-S-methyl, oxydemeton-methyl
	Cotton helopeltis	DDT
	Cowpea bruchids	(fumigation in stores usually)
	Maruca	Carbaryl, DDT, BHC
	Mexican bean beetle	Dieldrin, malathion
	Pea aphid	Dimethoate, formothion, demeton-S-methyl, menazon, diazinon, pyrethrum
	Pea pod borer	Carbaryl, tetrachlorvinphos, fenitrothion azinphos-methyl
	Pollen beetles	DDT
	Red cotton mite	Dimethoate, dicofol, tetradifon, etc.
	Spiny brown bugs	DDT, endosulfan
	Striped blister beetles	Dieldrin, parathion
Rice	African armyworm	DDT, endosulfan, malathion, trichlorphon, carbaryl
	Black paddy bug	DDT/BHC
	Dark-headed rice borer	Endrin, dieldrin, parathion, diazinon, trichlorphon, dichlorvos
	Green rice leafhopper	Carbaryl, malathion, azinphos-methyl, endosulfan
	Green stink bug	DDT, BHC, phorate
	Leaf-footed plant bug	BHC, parathion
	Lesser armyworm	DDT, BHC, endrin, parathion
	Paddy hispid	BHC, DDT, dieldrin, endrin, diazinon, phosphamidon, demeton-S-methyl
	Paddy stem borers	Endrin, dieldrin, parathion, diazinon, trichlorphon, dichlorvos
	Pink stalk borer	Endrin, dieldrin, parathion, diazinon, trichlorphon, dichlovos
	Purple stem borer	Endrin, dieldrin, parathion, diazinon, trichlorphon, dichlorvos
	Rice armyworms	DDT, BHC, endrin, parathion

	Rice bugs	Carbaryl, malathion, azinphos-methyl, endosulfan
	Rice caseworm	Parathion, BHC, dieldrin, malathion
	Rice cutworm	DDT, endosulfan, malathion, carbaryl, trichlorphon
	Rice hispid	BHC, DDT
	Rice leaf beetle	DDT, BHC, phosphamidon
	Rice leaf-miner	Dieldrin, heptachlor
	Rice leaf-roller	DDT, BHC, dieldrin, endosulfan, fenithion, fenitrothion, phosphamidon phosphamidon
	Rice shield bug	BHC, carbaryl, malathion, trichlorphon
	Rice skipper	Dieldrin, BHC, carbaryl
	Rice stalk borer	Endrin, dieldrin, parathion, diazinon, trichlorphon, dichlorvos
	Rice stem gall midge	BHC, endrin, dieldrin, dimethoate, diazinon, carbaryl, phorate, parathion, phosphamidon
	Rice thrips	DDT, BHC, malathion, diazinon, dimethoate
	Small brown planthopper	Carbaryl, malathion, azinphos-methyl, endosulfan
	White-backed planthopper	Aldrin, dieldrin, DDT, endrin, dimethoate, disulfoton, phosphamidon, carbaryl, malathion, endosulfan, azinphos-methyl
	White rice borer	Endrin, dieldrin, parathion, diazinon, trichlorphon, dichlorvos
	Zigzag-winged leafhopper	Carbaryl, malathion, azinphos-methyl, endosulfan
Rubber	Red cotton mite	Dimethoate, dicofol, tetradifon, etc.
Simsim (sesame)	Simsim gall midge	(not usually required)
	Simsim webworm	DDT
Sisal	Sisal weevil	Aldrin, dieldrin
Sorghum	African armyworm	DDT, endosulfan, malathion, carbaryl, trichlorphon
	American bollworm	DDT, BHC, DDT/BHC
	Blue bugs	DDT
	Coastal stalk borer	DDT
	Cotton lygus	DDT
	Epilachna beetle	Dieldrin, malathion
	Maize aphid	Dimethoate, demephion, demeton-S-

Crop	Pest	Insecticide
		methyl, formothion, menazon, ethoate-methyl
	Maize stalk borer	DDT
	Pink stalk borer	Endrin, dieldrin, parathion, diazinon, trichlorphon, dichlorvos
	Sorghum gall midge	DDT, carbaryl — not very effective
	Sorghum shoot fly	Endosulfan, phorate, disulfoton
	Spotted stalk borer	DDT, endrin
	Sugar cane stalk borer	Endrin
Sugar cane	Coastal stalk borer	DDT, endrin
	Pink stalk borer	Endrin, dieldrin, parathion, trichlorphon, diazinon, dichlorvos
	Pink sugar cane mealybug	(Dieldrin to control the ants)
	Sugar cane spittlebugs	BHC, dieldrin, phorate, toxaphene, carbofenothion
	Spotted stalk borer	DDT
	Sugar cane planthopper	BHC, dimethoate, dicrotophos, dimefox, parathion
	Sugar cane scale	Malathion, white oil
	Sugar cane stalk borer	Endrin
	Sugar cane termite	Dieldrin
	Sugar cane white grub	Aldrin, dieldrin, BHC, heptachlor
Sweet potato	Striped sweet potato weevil	(not usually required)
	Sweet potato clearwing	(not usually required)
	Sweet potato moth	TDE, parathion
	Sweet potato weevils	DDT
Tea	Black tea thrips	Fenitrothion
	Cotton helopeltis	DDT
	Red coffee mite	Dimethoate, dicofol
	Red crevice mite	Dicofol, chlorobenzilate
	Tea root weevil	DDT, dieldrin
	Yellow tea mite	Dicofol
Tobacco	American bollworm	DDT, BHC, DDT/BHC
	Black cutworm	DDT, dieldrin, carbaryl, fenitrothion, trichlorphon, tetrachlorvinphos
	Tobacco (onion) thrips	Dieldrin, diazinon, DDT, malathion
	Tobacco whitefly	DDT, dimethoate, pyrethrum
Tomato	American bollworms	DDT, BHC, DDT/BHC
	Green stink bug	DDT, BHC, phorate
	Tobacco (onion) thrips	DDT, dieldrin, diazinon, malathion
	Tobacco whitefly	DDT, dimethoate, pyrethrum

Wheat	Black wheat beetle	Aldrin, dieldrin, DDT, BHC, chlordane
	Epilachna beetle	Dieldrin, malathion
	Shiny cereal weevil	DDT
	Wheat aphid	Dimethoate, demephion, demeton-S-methyl, menazon, formothion, ethoate-methyl
Yam	Yam beetle	Aldrin, dieldrin, BHC, heptachlor

Table 16.3 Fungicide usage

Diseases controlled / **Chemical**	***Bacteria***	***Phycomycetes***	***Basidiomycetes***	***Sclerotinia* incl. *Botrytis***	***Powdery mildews***	***Ascomycetes, etc****	**Miscellaneous**
Bordeaux mixture	•	•	•	•		•	
Burgundy mixture	•	•	•	•		•	
Cheshunt compound							Seed beds and boxes
Copper(II) hydroxide	•	•	•	•		•	
Copper naphthenates							Wood preservative
copper oxychloride	•	•	•	•	(•)	•	
Copper(II) sulphate(VI)	•	•	•	•		•	Wood preservative, algicide and in Bordeaux mixtures
Copper(I) oxide	•	•	•	•	(•)	•	
Lime sulphur					•	•	
Oxine copper	•	•	•	•		•	Mainly used in soil or on utensils
Sulphur					•	•	
Mercury(I) and mercury(II) chlorides	•	•	•	•	•	•	Used on soil or turf or as a root dip
Methoxymethyl mercury salts	•	•	•	•	•	•	Primarily as seed dressing
Phenylmercury acetate and chloride	•	•	•	•	•	•	Primarily as seed dressing
Other organo-mercurials	•	•	•	•	•	•	Primarily as seed dressing
Triphenyltin compounds		•	•	•	•	•	
Methyl arsenic sulphide			•	•	•	•	Primarily as seed dressing
Cufraneb	•	•	•	•		•	
Ferbam		•	•	•		•	
Mancozeb		•	•	•		•	
Maneb		•	•	•		•	
Metiram		•	•	•		•	

Propineb		•	•	•	•	•	
Thiram		•	•	•		•	Used especially in seed dressings
Zineb		•	•	•		•	
Ziram		•	•	•		•	
Anilazine		•	•	•		•	
Captafol		•		•		•	
Captan		•	•	•		•	
Chlorthalonil		•	•	•	•	•	
Dichlorfluanid		•		•	•	•	
Ditalimfos					•	•	
Dithianon	•	•	•			•	
Dodine				•	(•)	•	
Drazoxolon		•	•	•	•	•	
Fenarimol					•	•	
Folpet		•	•	•	•	•	
Halacrinate			•		•	•	
Prochloraz					•	•	
Pyridinitril		•				•	
Tolylfluanid		•		•	•	•	
Bronopol	•						
Chloranil				•		•	
Dichlone		•		•		•	Algicide
Etridiazole		•					
Fenaminosulf		•				•	
Fenfuram			•	•		•	
Guazatine			•			•	
Hexachlorobenzene			(•)				
Hydroxyisoxazole		•	•			•	
Quinacetol sulphate						•	Potato tubers
Quintozene			•	•		(•)	Not effective against *Fusarium*
Binapacryl					•		Acaricide
Chlorquinox					•		
Dicloran		•		•		•	Also for post-harvest fruit rots
Dinobuton					•		Acaricide
Dinocap					•		Acaricide
Edifenphos			•			•	Rice diseases
Fluotrimazole					•		
Iprodione			•	•		•	
Nitrothal-isopropyl					•		
Petroleum oils						(•)	Sigatoka disease of bananas
Phenylphenol				•		•	Post-harvest use
Procymidone				•			

Quinomethionate				•	•	Acaricide
Tecnazene					(•)	Fusarium rot of potatoes
Thioquinox				•		
Vinclozolin			•			
Aluminium *tris* (ethyl phosphonate)	•					
Benodanil		•				
Benomyl		(•)	•	•	•	
Bupirimate				•		
Carbendazim		(•)	•	•	•	
Carboxin		•				Seed treatment
Chloroneb	•	•			•	Seed and soil application
Diclobutrazol		•		•	•	
Dimethirimol				•		
Dodemorph		•	•	•	•	
Ethirimol					•	Especially used as seed dressing
Fuberidazole		•			•	
Furalaxyl	•					
IBP		•			•	Used on rice
Imazadil	•			•	•	
Isoprothiolane		•			•	Used on rice
Metalaxyl	•					
Oxycarboxin		•				
Prothiocarb	•					
Pyrocarbolid		•				
Pyrazophos				•		
Thiabendazole	•	(•)	•	•	•	
Thiophanate and Thiophanate methyl			•	•	•	
Triadimefon		•		•	•	
Tricyclazole					•	Rice blast
Tridemorph				•	•	
Triforine		•		•	•	

Notes

* *includes most Fungi Imperfecti except those with Oomycete or Basidiomycete perfect states*

(•) partially effective against some members of the groups.

Further reading

Jeffs, K.A. (ed.) (1978). *Seed treatment.* CIPAC Monograph No. 2.
MAFF (1961). *Farm sprayers and their uses*. Bull. No. 182, HMSO: London.
Matthews, G.A. (1979). *Pesticide application methods*. Longman: London.
Mulder, D. (ed). (1979). *Soil disinfestation.* Elsevier: Amsterdam.

Part 4 Chemicals

17 Materials used to control plant pests

Chlorinated hydrocarbons

These compounds are often referred to as the **organochlorines**. They are a broad spectrum and very persistent group, which usually kill both by contact and as stomach poisons. Generally, they are more effective against insects with biting mouthparts than the sap-suckers.

Because of their persistence they get taken up in food chains very easily and accumulate in the body fat of the vertebrate predators at the apex of the food chains. Under normal conditions this build-up in the body fat may not affect the animal. It is in times of starvation when the body fat is broken down that the amount of pesticide released into the blood may be critical or even fatal.

Most countries in Europe and North America are in the process of restricting the use of chlorinated hydrocarbons, where suitable alternatives are available, because of the long-term contamination dangers to the environment. It can also be observed that in most of these countries the major pests have already become or are now becoming resistant to the organochlorine compounds. In some cases suitable alternatives are not available, but in general their use is declining. However, most countries still use DDT, dieldrin, etc., against some insect pests for which no suitable alternative has yet been found.

The organochlorine compounds may be subdivided into several groups. The three most common are represented by DDT, gamma-HCH(BHC) and aldrin (cyclodienes). Despite the structural differences between the sub-groups they do possess several characteristics in common — they are chemically stable, have a low solubility in water, moderate solubility in organic solvents and liquids, and a low vapour pressure. The whole class are similar enough to produce similar physiological responses in the insects. The stability and solubility of the chemicals make most of the group very persistent.

The solubility of DDT in liquids enables the poison to penetrate the insect integument quite readily (as opposed to slow penetration through mammalian skin). It is this difference in penetrative ability which accounts for its selective toxicity to insects. Cuticle thickness does not seem greatly to influence the susceptibility of insects to DDT, although DDT may penetrate more readily through the flexible intersegmental membranes. Dissolution in the epicuticular wax is apparently essential to toxic action. Penetration rate of the poison increases with temperature, often nearly doubling for a 20 °C rise in temperature. The precise mode of action of these poisons on the insect is not at all fully understood as yet.

It must be stressed that in both Volume 1 and Volume 2 we make reference to the chemicals known to be effective in controlling a particular pest, or in a few cases pesticides that are likely to be effective. We have not withdrawn all references to chemicals such as the organochlorines (DDT, dieldrin, etc.) and some selected very toxic organophosphorus compounds such as parathion, just because their use is restricted in some countries. We are concerned with documenting the chemicals known to be effective against particular pests. The actual choice of a pesticide to be used in any local situation must be done in consultation with local recommendations which can be expected to vary quite considerably from country to country. For example DDT and dieldrin are still officially recommended for use against some pests in many countries, but in some countries they are totally banned.

Aldrin
(1,2,3,4,10,10-hexachloro-1,4,4a,5,8,8a-hexahydro-*exo*-1 4,-*endo*-5,8-dimethanonaphthalene)

Trade names 'Alrdine', 'Aldrex', etc.
Properties A broad-spectrum, persistent, non-systemic, non-phytotoxic insecticide with high

contact and stomach activity, effective against soil insects at rates of 0·6–5·6 kg per hectare.

Aldrin is stable to heat, alkali and mild acids, but the unchlorinated ring is attacked by oxidising agents and strong acids. It is oxidised to dieldrin. It is compatible with most pesticides and fertilisers. On storage, it slowly produces hydrogen chloride gas (HCl) which is corrosive.

The technical product is a brown colourless solid, practically insoluble in water but quite soluble in mineral oils, and readily soluble in propanone (acetone), benzene, and 1,2-dimethylbenzene (xylene).

The acute oral LD50 for rats is 67 mg per kilogram. It is absorbed through the skin.

Use Effective against all soil insects, e.g. termites, beetle adults and larvae, fly larvae, cutworms, crickets, etc.

Caution

1 Harmful to fish.
2 Treated seeds should not be used for human or animal consumption.
3 Avoid excessive skin contact.
4 Risks to wildlife are considerable on a long-term basis.

Formulations 30 per cent e.c.; 2·5–5 per cent dust; m.1.; 5 and 20 per cent granules; 20–50 per cent w.p.; liquid seed dressing; as an insecticidal lacquer.

DDT
(1,1,1-trichloro-2,2-di(chlorophenyl)ethane)

Properties A broad-spectrum stomach and contact poison, of high persistence, non-systemic and non-phytotoxic except to cucurbits.

The pp¹-isomer forms colourless crystals, practically insoluble in water, moderately soluble in petroleum oils and readily soluble in most aromatic and chlorinated solvents. The technical product is a waxy solid. DDT is dehydrochlorinated at temperatures above 50 °C, a reaction catalysed by ultraviolet light. In solution it is readily dehydrochlorinated by alkalis or organic bases. It is stable to acid and alkaline manganate(VII) and to aqueous acids and alkalis.

The acute LD50 for male rats is 113 mg per kilogram. DDT is stored in the body fat of birds and mammals and excreted in the milk of mammals.

Use Effective against most insects, but with little action on phytophagous mites.

Caution

1 Harmful to bees, fish and livestock.
2 Pre-harvest interval for edible crops — 2 weeks.
3 Do not use on cucurbits or certain barley varieties, as damage may occur.

Formulation 25 per cent e.c.; 25 per cent m.1.; 50 per cent w.p.; 5 per cent dusts; smokes. May be combined with gamma-HCH or malathion in sprays or smokes.

Dicofol
(2,2,2-trichloro-1,1-di-(4-chlorophenyl)ethanol)

Trade name 'Kelthane'

Properties It is a non-systemic acaricide, with little insecticidal activity, recommended for the control of mites on a wide range of crops. Although residues in soil decrease rapidly, traces may remain for one year or more.

The pure compound is a white solid, practically insoluble in water, but soluble in most aliphatic and aromatic solvents. It is hydrolysed by alkali, but it is compatible with all but highly alkaline pesticides. Wettable powder formulations are sensitive to solvents and surfactants, which may affect acaricidal activity and phytotoxicity.

The acute oral LD50 for male rats is 809 ± 33 mg per kilogram.

Use Effectively only against Acarina; kills eggs and all active stages of the mites.

Caution

1 Pre-harvest interval for edible crops is 2–7 days.
2 Safe to bees.

Formulation 18·5 per cent and 42 per cent e.c.; 30 per cent dust.

Dieldrin
(1,2,3,4,10,10-hexachloro-6,7-epoxy-1,4,4a,5,6,7,8,8a-octahydro-*exo*-1,4-*endo*-5,8-dimethanonaphthalene)

Properties A broad-spectrum, persistent, non-systemic, non-phytotoxic insecticide made by oxidation of aldrin. It is of high contact and stomach activity.

It is stable to light, alkali and mild acids, and it is compatible with most other pesticides. Dieldrin occurs as white odourless crystals. The technical product consists of light brown flakes, practically insoluble in water, slightly soluble in petroleum oils, moderately soluble in propanone, soluble in aromatic solvents.

The acute oral LD50 for male rats is 46 mg per kilogram. It can be absorbed through the skin.

Use Effective against most insects.

Caution

1 Harmful to fish.
2 Safe to bees.
3 Seed dressings containing organomercury compounds can cause rashes or blisters on the skin. Those containing thiram can be irritating to the skin, eyes, nose and mouth.
4 Treated seed should not be used for human or animal consumption.
5 Dressed seeds are dangerous to birds.

Formulation 15 per cent e.c.; m.1; 50 per cent w.p.; dry seed dressings. Some seed dressings made with thiram or organomercury compounds.

Endosulfan
(6,7,8,9,10,10-hexachloro-1,5,5a,6,9,9a-hexa-hydro-6,9,-methano-2,4,3 benzo(e)dioxathiepin 3-oxide)

Trade names 'Thiodan', 'Cyclodan'

Properties It is a non-systemic contact and stomach insecticide and acaricide. The technical product is a brownish crystalline solid, practically insoluble in water, but moderately soluble in most organic solvents. It is a mixture of 2 isomers. It is stable to sunlight but subject to slow hydrolysis to the alcohol and sulphur dioxide. It is compatible with non-alkaline pesticides.

The acute oral LD for rats is 110 (55 - 220) mg per kilogram.

Use Effective against most crop mites and some Hemiptera (aphids, capsids) and some beetles.

Caution

1 This is a poisonous substance (Part II p.93) — full protective clothing should be worn.
2 Extremely dangerous to fish.
3 Dangerous to livestock.
4 Harmful to bees.
5 Pre-harvest interval for fruit is 6 weeks.
6 Pre-access interval to treated areas is, for unprotected persons, 1 day; for animals and poultry 3 weeks.

Formulation 17·5 per cent and 35 per cent e.c.; 17·5 per cent, 35 and 50 per cent w.p.; 1, 3, 4 and 5 per cent dusts and 5 per cent granules.

Endrin
(1,2,3,4,10,10-hexachloro-6,7-epoxy-1,4,4a,5,6,7,8,8a-octahydro-*exo*-1,4-*endo*-5,8-dimethanonaphthalene)

Trade name 'Endrex'

Properties A very toxic, broad-spectrum, persistent insecticide and acaricide, isomeric with dieldrin. It is non-systemic, and non-phytotoxic at insecticidal concentrations, but may damage maize.

It is a white crystalline solid practically insoluble in water, sparingly soluble in alcohols and petroleum oils, moderately soluble in benzene and propanone. The technical product is a light brown powder of not less than 85 per cent purity. It is stable to alkali and acids but strong acids or heating above 200 °C cause a rearrangement to a less insecticidal derivative. It is compatible with other pesticides.

The acute oral LD for male rats is 17·5 mg per kilogram.

Use Effective against many insects and mites, used mainly on field crops.

Caution

1 A very poisonous pesticide (Part II p. 93) — full protective clothing should be worn.
2 Dangerous to bees, fish, livestock, wild birds and animals.
3 Fruit should not be sprayed after flowering.
4 Minimum interval to be observed between last application and access to treated areas is, for unprotected persons, 1 day; for animals and poultry, 3 weeks.

Formulation 20 per cent liquid formulation; or dust.

Gamma-HCH or gamma-BHC (1,2,3,4,5,6-hexachlorocyclohexane)

Trade names 'Lindane', 'Gammexane', etc.

Properties Gamma-HCH (benzene hexachloride) exists as 5 isomers in the technical form but the active ingredient is the gamma-isomer. 'Lindane' is required to contain not less than 99 per cent gamma-HCH. It exhibits a strong stomach poison action, persistent contact toxicity and fumigant action against a wide range of insects. It is non-phytotoxic at insecticidal concentrations. The technical HCH causes 'tainting' of many crops but there is less risk of this with 'Lindane'

'Lindane' is stable to air, light, heat and carbon dioxide; unattacked by strong acids, but can be dehydrochlorinated by alkalis.

It occurs as colourless crystals and is practically insoluble in water; slightly soluble in petroleum oils; soluble in propanone, aromatic and chlorinated solvents.

The acute oral LD50 for rats is 88 mg per kilogram.

Use Effective against many soil insects, e.g. beetle adults and larvae, fly larvae, *Collembola* spp., and also against many other biting and sucking insects, e.g. aphids, psyllids, whiteflies, capsids, midges, sawflies and thrips.

Caution

1 Dangerous to bees.
2 Harmful to fish and livestock.
3 Pre-harvest interval about 2 weeks.
4 Treated seeds should not be used for human or animal consumption.

Formulation Many different seed dressings, some with mild organomercury compounds or with captan or thiram; 50 per cent w.p. with bran as bait; dust; liquid e.c. or suspension.

Heptachlor (1,4,5,6,7,8,8-heptachloro-3a,4,7,7a-tetrahydro-4,7--methanoindene(I))

Properties A broad-spectrum, non-systemic, stomach and contact insecticide with some fumigant action.

It is a white crystalline solid, practically insoluble in water, slightly soluble in alcohol, more so in kerosene; it is stable to light, moisture, air and moderate heat. It is compatible with most pesticides and fertilisers.

The acute oral LD50 for male rats is 100 mg per kilogram.

Use Effective against many different insect species.

Formulation As e.c.; w.p.; dusts and granules of various a.i. contents.

Mirex (dodecachlorooctahydro-1,3,4-metheno-1*H*-cyclobuta(*cd*)pentalene)

Properties It is a stomach insecticide, with little contact effect, used mainly against ants.

It is a white solid of negligible volatility. Insoluble in water but moderately soluble in benzene, tetrachloromethane and 1,2-dimethylbenzene.

Acute LD50 for male rats is 306 mg per kilogram.

Use Mostly used in baits against fire ants and harvester ants.

Formulation Used at 1·5 or 75 g a.i. per kg.

TDE (1,1-dichloro-2,2-di-(4-chlorophenyl)ethane)

Trade name 'DDD'

Properties It is a non-systemic contact and stomach insecticide which has less potency than DDT in general, but is of equal or greater potency against certain insects, e.g. leaf-rollers, mosquito larvae, hornworms. It is non-phytotoxic except possibly to cucurbits.

The pure compound forms colourless crystals practically insoluble in water but soluble in most organic compounds. Its chemical properties resemble those of DDT but it is more slowly hydrolysed by alkali.

The acute oral LD50 for rats is 3 400 mg per kilogram.

Use Particularly effective against leaf-rollers,

mosquito larvae, hornworms; effective against many caterpillars, weevils, thrips and earwigs.

Caution

1 Harmful to bees, fish and livestock.
2 Pre-harvest interval for edible crops in 2 weeks.
3 Pre-access interval for livestock to treated areas is 2 weeks.

Formulation 50 per cent w.p.; e.c. 25 per cent; 5 and 10 per cent dusts.

Tetradifon
(4-chlorophenyl 2,4,5-trichlorphenylsulphone)

Trade names 'Tedion V-18'

Properties A systemic acaricide toxic to the eggs and all stages of phytophagous mites except adults. At acaricidal concentrations it is non-phytotoxic.

It forms colourless crystals, almost insoluble in water, slightly soluble in alcohols and propanone, more soluble in aromatic hydrocarbons and trichloromethane. It is resistant to hydrolysis by acid or alkali. It is compatible with other pesticides and is non-corrosive.

The acute oral LD 50 for rats is more than 5 000 g per kilogram

Use Effective against eggs, larvae, nymphs of phytophagous mites, but not adults. Recommended for application to top fruit, citrus, tea, cotton, grapes, vegetables, ornamentals and nursery stock.

Caution

1 Do not use smokes to young cucumbers or plants that are wet or damage may occur.
2 At correct dosage is harmless to beneficial insects.
3 Safe to bees.

Formulation 20 per cent w.p.; 18 per cent e.c.; may be combined with malathion in smoke formulations.

Tetrasul
(4-chlorophenyl 2,4,5-trichlorphenylsulphide)

Trade name 'Animert V-101'

Properties A non-systemic acaricide, highly toxic to eggs and all stages of phytophagous mites except adults. At the correct dosage it is non-phytotoxic. As it is highly selective it does not pose a hazard to beneficial insects or wildlife.

It is a brown crystalline solid, only slightly soluble in water, moderately soluble in propanone and ethoxyethane (ether) but soluble in benzene and trichloromethane. Stable under normal conditions, but should be protected against prolonged exposure to sunlight; it is oxidised to its sulphone, tetradifon. It is non-corrosive, and compatible with most other pesticides.

The acute oral LD50 for female rats is 6 810 mg per kilogram.

Use Effective against eggs, larvae and nymphs of most phytophagous mites, but not adults. Recommended for use on fruit and cucurbits at the time when the winter eggs are hatching.

Formulation 18 per cent e.c. and 18 per cent w.p.

Substituted phenols

The nitrophenols are not likely to compete with the newer insecticides; they are used mainly as herbicides and for the control of powdery mildews. The dinitrophenols have mammalian toxicity so high as to restrict their usefulness in crop protection.

Amitraz
(*NN*-di-(2,4-xylyliminomethyl)-methylamine)

Trade names 'Mitac', 'Taktic', 'Triatox'

Properties An acaricide effective against a wide range of phytophagous mites; all stages are susceptible. It has some insecticidal properties and is effective against some bugs and lepidopteran eggs. Relatively non-toxic to bees and predatory insects.

It is insoluble in water, but soluble in propanone and methylbenzene. At acid pH levels it is unstable, and will slowly deteriorate if moist. Compatible with most commonly-used pesticides.

The acute oral LD50 for rats is 800 mg per kilogram.

Use Effective against phytophagous mites, especially red spider mites, at concentrations from 20–50 g a.i. per 100 litres, depending on the species. Also effective against eggs of *Heliothis*. Used against ticks and mites of cattle and sheep.

Caution

1 This is a poisonous substance (Part II p. 93) — protective clothing should be worn.
2 Pre-harvest interval for fruit is 2 weeks.
3 Harmful to fish.

Formulation e.c. 200 g a.i. per litre; d.p. 250 and 500 g a.i. per kilogram.

Binapacryl
(2-*sec*butyl-4,6-dinitrophenyl 3-methylcrotonate (I))

Trade names 'Morocide', 'Acricid', 'Endosan'

Properties It is a non-systemic acaricide (also effective against powdery mildews) and mainly used against red spider mites. Non-phytotoxic to a wide range of apples, pears, cotton and citrus; some risk of damage to young tomatoes, grapes and roses.

It is a white crystalline powder practically insoluble in water but soluble in most organic solvents. It is unstable in concentrated acids and dilute alkalis, suffers slight hydrolysis on long contact with water and is slowly decomposed by ultraviolet light. It is non-corrosive, and compatible with w.p. formulations of insecticides and non-alkaline fungicides. With organophosphorus compounds it may be phytotoxic.

The acute oral LD50 for rats is 120 - 165 mg per kilogram.

Use Effective especially against red spider mites.

Caution

1 Harmful to fish and livestock.
2 Safe to bees.
3 Pre-harvest interval for edible crops is 1 week.
4 Pre-access interval for livestock to treated areas is 4 weeks.

Formulation 25 and 50 per cent w.p.; 40 per cent e.c.; 4 per cent dust.

DNOC
(4,6-dinitro-o-cresol (I))

Trade name 'Sinox'

Properties It is a non-systemic stomach poison and contact insecticide; ovicidal to the eggs of certain insects. It is strongly phytotoxic and its use as an insecticide is limited to dormant sprays or on waste ground, e.g. against locusts. It is also used (not usually in oil) as a contact herbicide for the control of broad-leaved weeds in cereals. And in e.c. formulations for the pre-harvest desiccation of potatoes and leguminous seed crops.

It forms yellowish, colourless crystals, only sparingly soluble in water, but soluble in most organic solvents and in ethanoic (acetic) acid. The alkali salts are water soluble. It is explosive, and usually moistened with up to 10 per cent water to reduce the hazard; it is corrosive to mild steel in the presence of water.

Use Effective against overwintering stages of aphids, capsids, psyllids, scale insects, red spider mites and various Lepidoptera on top, bush and cane fruit. Products containing DDT also control various weevils and tortrix moths in their overwintering stages. Also used as herbicide.

Caution

1 If the concentrated substance contains more than 5 per cent of DNOC it is a Part II (p. 93) substance and protective clothing should be worn.
2 Dangerous to fish.
3 Very phytotoxic; use only as a dormant spray or on waste ground.

Formulation The insecticide formulation is an e.c. in petroleum oil; and formulated with DDT.

Pentachlorophenol

Trade names 'Dowicide 7', 'Santophen 20'.

Properties An insecticide used for termite control; a fungicide used for protection of timber from fungal rots and wood-boring insects. It is strongly phytotoxic, and is used as a pre-harvest defoliant and as a general herbicide.

It forms colourless crystals which are volatile in steam, almost insoluble in water and soluble in most organic solvents. It is non-corrosive in the absence of moisture, solutions in oil cause deterioration of natural rubber but synthetic rubbers may be used in equipment and protective clothing.

The acute oral LD50 for rats is 210 mg per kilogram; it irritates mucous membranes and causes sneezing. The solid, and aqueous solutions stronger than 1 per cent, cause skin irritation.

Use Effective against termites and other wood-boring insects; use restricted by strong phytotoxicity.

Caution

1 Very phytotoxic.
2 Irritation to skin and mucous membranes.

Formulation Used undiluted or formulated in oil. 'Santobrite' and 'Dowicide G' are the technical sodium salt.

Organophosphorus compounds

These were discovered and developed during the World War II by a German research team responsible for developing nerve gases; they rank among the most toxic substances known to man.

These compounds have phosphorus chemically bonded to the carbon atoms of organic radicals, and are effective as both contact and systemic insecticides and acaricides. Nearly 50 compounds are in current use against insects and mites. Many of these compounds are very toxic to mammals and birds and have to be handled with care. Doses may be accumulative. The systemic compounds are very effective against sap-sucking insects. All the organophosphorus compounds are relatively transient and are soon broken down to become non-toxic. Great care in timing and application is required in the use of these compounds for effective results. The mode of action in both insects and mammals appears to be inhibition of the enzyme acetylcholinesterase.

Azinphos-methyl
(*S*-(3,4-dihydro-4-oxobenzo[*d*]-[1,2,3]-triazin-3-ylmethyl)

Trade names 'Gusathion M', 'Guthion'

Properties A non-systemic broad-spectrum insecticide and acaricide of relatively long persistence, with contact and stomach action. It forms white crystals, almost insoluble in water but soluble in most organic solvents. It is unstable at temperatures above 200 °C and is rapidly hydrolysed by cold alkali and acid.

The acute oral LD50 for rats is 16·4 mg per kilogram.

Use Effective against Lepidoptera, mites, aphids, whiteflies, leafhoppers, scales, thrips, grasshoppers, some fly larvae and some beetles.

Caution

1 A poisonous substance (Part II p. 93) — protective clothing should be worn.
2 Dangerous to bees.
3 Harmful to fish and livestock.
4 Pre-harvest interval for edible crops is 4 weeks according to crop and country.
5 Pre-access interval for livestock to be treated is 2 weeks.

Formulation 20 per cent e.c.; 25 and 50 per cent w.p.; 2·5 and 5 per cent dusts; and ULV formulations.

Azinphos-methyl with demeton-S-methyl sulphone
(*S*-2-ethylsulphonylethyl *OO*-dimethyl phosphorothioate)

Trade name 'Gusathion MS'

Properties This mixture combines the properties of both pesticides — usually in the proportion of 75 – 25 per cent. It practice, the mixture acts like azinphos-methyl but with a systemic action.

Use Effective against thrips, aphids and fly larvae (midges).

Bromophos
(*O*-(4-bromid-2,5-dichloro-phenyl) *OO*-dimethyl phosphorothioate)

Trade name 'Nexion'

Properties A broad-spectrum contact and stomach insecticide. Non-phytotoxic at insecticidal concentrations, but suspect of damage under glass. Persists on sprayed foliage for 7 – 10 days. No systemic action.

It occurs as yellow crystals, relatively insoluble in water, but soluble in most organic solvents, particularly tetrachloromethane, ethoxyethane (ether) and methylbenzene. It is stable in media up to pH 9·0, non-corrosive and compatible with all pesticides except sulphur and the organometal fungicides.

The acute LD50 for rats is 3 750 - 7 700 mg per kilogram.

Use Effective against a wide range of insects on crops at concentrations of 27 - 75 a.i. per 100 litres.

Caution

1 Harmful to fish and bees.
2 Possibly phytotoxic under glass.

Formulation e.c. 250, 400 g a.i. per litre; w.p. 250 g a.i. per kilogram; dusts 20 - 50 g a.i. per kilogram; atomising concentrate 400 g a.i. per litre; coarse powder 30 g a.i. per kilogram; granules 50 - 100 g a.i. per kilogram.

Bromopropylate
(isoprophyl 4,4-dibromobensilate (1))

Trade names 'Acarol', 'Neoron'

Properties A contact acaricide with residual activity, useful for many crops.

It is a crystalline solid, insoluble in water but readily soluble in most organic solvents; stable in neutral media.

The acute oral LD50 for rats is 5 000 mg per kilogram.

Use Used against mites on pome and stone fruits, citrus, hops, cotton, beans, cucurbits, tomatoes, strawberries and ornamentals.

Formulation As 500 and 250 g per litre.

Carbophenothion
(*S*-4-chlorophenylthiomethyl *OO*-diethyl phosphorodithioate)

Trade names 'Trithion', 'Garrathion'

Properties A non-systemic acaricide and insecticide, with a long residual action. It is phytotoxic at high concentration to some plants. Fifty per cent degradation in soil occurs in 100 days or longer, depending upon the soil type.

It is a pale amber liquid; insoluble in water, but miscible with most organic solvents. Relatively stable to hydrolysis but oxidised on the leaf surface to the phosphorothiolate. It is compatible with most pesticides, and is non-corrosive to mild steel.

The acute oral LD50 for male rats is 32·2 mg per kilogram.

Use Used mainly on decidous fruit, in combination with petroleum oil, as a dormant spray for the control of overwintering mites, aphids and scale insects; on citrus as an acaricide.

Caution Harmful to livestock, birds and wild animals.

Formulation e.c. 0·2, 0·4, 0·6, 0·8 kg per litre; 25 per cent w.p.; dusts of 1, 2 and 3 per cent.

Chlorfenvinphos
(2-chloro-1-(2,4-dichlorophenyl)vinyl diethyl phosphate)

Trade names 'Birlane', Sapecron'

Properties A relatively short-lived insecticide, effective against soil insects, non-phytotoxic at recommended dosages. The pure compound is an amber-coloured liquid, sparingly soluble in water but miscible with propanone, ethanol, kerosene and 1,2-dimethylbenzene. It is stable when stored in glass or polythene vessels, but is slowly hydrolysed by water. It may corrode iron and brass on prolonged contact and the e.c formulations are corrosive to tin plate.

Acute oral LD50 for rats is 10 - 39 mg per kilogram.

Use Particularly effective against root flies, root worms and cutworms as soil applications. As a foliage insecticide is recommended for the control of Colorado beetle on potato, leafhoppers on rice, and stem-borers on maize, sugar cane and rice. The half-life in soil is normally only a few weeks.

Caution

1 This is a poisonous substance (Part II p. 93) — protective clothing should be worn.
2 Dangerous to fish.
3 Use seed dressings carefully to avoid risks to birds.
4 Treated seed should not be used for human or animal consumption.
5 Pre-harvest intervals for edible crops — 3 weeks.

Formulation 24 per cent e.c; 25 per cent w.p; 5 per cent dust; 10 per cent granules; seed dressings (liquid) 40 per cent (+ 2 per cent mercury compounds).

Chlorpyrifos
(*00*-diethyl 0-3,5,6-trichloro-2-pyridyl phosphorothioate)

Trade name 'Dursban', 'Lorsban'

Properties A broad-spectrum insecticide with contact, stomach and vapour action. It has no systemic action. It is volatile enough to make insecticidal deposits on nearby untreated surfaces. At insecticidal concentrations it is non-phytotoxic. It persists in soil for 2 – 4 months.

It forms white crystals; insoluble in water, but soluble in methanol and most other organic solvents. Stable under normal storage conditions. It is compatible with non-alkaline pesticides, but is corrosive to copper and brass.

The acute oral LD50 for male rats is 163 mg per kilogram.

Use Effective against various soil and many foliar insect and mite pests; also for flies, mosquitoes and household pests, and ectoparasites of cattle and sheep. Specifically used for control of root maggots, aphids, capsid bugs, caterpillars and red spider mites.

Caution
1 Dangerous to bees, fish and shrimps.
2 Pre-harvest interval for all edible crops is 2 – 6 weeks.

Formulation w.p. 25 per cent; e.c. 0·2 and 0·4 kg per litre; granules 1 – 10 per cent.

Demephion
(*00*-dimethyl *0*-2-methylthioethyl phosphorothioate (I) and *00*-dimethyl *S*-2-methylthioethyl phosphorothioate (II))

Trade name 'Cymetox'

Properties A systemic insecticide and acaricide effective against sap-feeding insects and non-phytotoxic to most crops.

It is a straw-coloured liquid (a mixture of 2 isomers), miscible with most aromatic solvents, chlorobenzene and ketones, immiscible with most aliphatic solvents. It is generally non-corrosive and compatible with most, except strongly alkaline, pesticides.

The acute oral LD50 for rats is about 0·15 ml per kilogram; the acute dermal LD50 is about 0·6 ml per kilogram.

Use Mainly used against aphids, on all crops.

Caution
1 This is a poisonous substance (Part II p. 93) — protective clothing should be worn.
2 Harmful to bees, livestock, fish, game, wild birds and mammals.
3 Pre-harvest interval for edible crops is 3 weeks.
4 Pre-access interval for livestock to treated areas is 2 weeks.

Formulation 30 per cent e.c.

Demeton
(*00*-diethyl *0*-2-ethylthioethyl phosphorothioate

Trade name 'Systox'

Properties A systemic insecticide and acaricide with some fumigant action, effective especially against sap-sucking insects and mites. No marked phytotoxicity has been recorded.

The technical product is a light yellow oil, hydrolysed by strong alkali but is compatible with most alkaline pesticides. It is almost insoluble in water but is soluble in most organic solvents.

The acute oral LD50 for male rats is 30 mg per kilogram.

Use Effective against sap-sucking insects and mites.

Caution Harmful to bees, fish and livestock.

Formulation e.c. of different oil contents.

Demeton-S-methyl
(*S*-2-ethylthioethyl *00*-dimethyl phosphorothioate)

Trade name 'Metasystox (i),

Properties A systemic and contact insecticide and acaricide, metabolised in the plant to the sulphoxide and sulphone. It is rapid in action with moderate persistence.

It is a colourless oil, only slightly soluble in water, but soluble in most organic compounds. It is hydrolysed by alkali.

The acute oral LD50 for rats is 65 mg per kilogram of the technical material or 40 mg per kilogram of the pure substance.

Use Effective against most sap-sucking pests (aphids, leafhoppers etc.), sawflies and red spider mites.

Caution

1 This is a poisonous substance (Part II p. 93) — protective clothing should be worn.
2 Certain ornamentals, especially some chrysanthemums, may be damaged by sprays.
3 Harmful to bees, fish, livestock, game, wild birds and mammals.
4 Pre-harvest interval for edible crops is 2–3 weeks.
5 Pre-access interval of livestock to treated areas is 2 weeks.

Formulation 25 and 50 per cent e.c. with emulsifier chosen to reduce dermal hazards.

Diazinon
(*OO*-diethyl *O*–2-isopropyl-6-methylpyrimidin-4-yl phosphorothioate)

Trade name 'Basudin'

Properties A non-systemic insecticide with some acaricidal action, used mainly against flies, both in agricultural and veterinary practice. At higher dosages it may be phytotoxic.

It is a colourless oil almost insoluble in water, but is miscible with ethanol, propanone, 1,2-dimethylbenzene, and is soluble in petroleum oils. It decomposes above 120 °C and is susceptible to oxidation; stable in alkaline media but is slowly hydrolysed by water and dilute acids. It is compatible with most pesticides but should not be compounded with copper fungicides.

The acute oral LD50 for male rats is 108 mg per kilogram.

Use Especially effective against flies and their larvae, e.g. Antomyiidae on vegetables and carrot fly, also used against mites, thrips, springtails, glass house pests and some bugs (aphids, capsids).

Caution

1 Dangerous to bees.
2 Harmful to fish, livestock, game, wild birds and mammals.
3 Overdosage may lead to phytotoxicity on some crops.
4 Pre-harvest interval for edible crops is usually 2 weeks.
5 Pre-access interval for livestock to treated areas is 2 weeks.

Formulation Aerosol solutions; 25 per cent e.c.; 40 and 25 per cent w.p.; 4 per cent dust; 5 per cent granules.

Dichlorvos
(2,2-dichlorovinyl dimethyl phosphate (I))

Trade names 'Vapona', 'Nogos', 'Oko', 'Mafu', 'Dedevap', 'Nuva'

Properties A short-lived contact and stomach insecticide with fumigant and penetrant action, non-phytotoxic. It is used as a household and public-health fumigant, especially against mosquitoes and other Diptera, and for crop protection.

It is barely soluble in water but is miscible with most organic solvents and aerosol propellants. It is a colourless to amber liquid, stable to heat, but is easily hydrolysed; corrosive to iron and mild steel but non-corrosive to stainless steel and aluminium.

The acute oral LD50 for male rats is 80 mg per kilogram.

Use Especially effective against flies, but also kills most other crop pests; often used for glasshouse fumigation as it kills most glasshouse pests. Also used on outdoor fruit and vegetables where rapid kill is required close to harvest. It will kill sap-sucking and leaf-mining insects.

Caution

1 A poisonous substance (Part II p. 93) requiring protective clothing to be worn.
2 Dangerous to bees.
3 Pre-harvest interval is 1 day.
4 Pre-access interval to treated areas is 12 hours.

Formulation 50 and 100 per cent e.c.; 'Vapona Pest Strip'; 0·4 to 1·0 per cent aerosols; 0·5 per cent granules.

Dimefox
(tetramethylphosphorodiamidic fluoride)

Trade name 'Terra Sytam'

Properties A systemic insecticide and acaricide of very high toxicity, and mainly for soil treatment of hops against aphids and red spider mites; non-phytotoxic at insecticidal concentrations.

It is a colourless liquid, miscible with water and most organic solvents. It is resistant to hydrolysis by alkali but is hydrolysed by acids, slowly oxidised by vigorous oxidising agents, rapidly by chlorine. Hence for decontamination treat with acids followed by bleaching powder. It is compatible with other pesticides, but the technical product slowly attacks metals.

The acute oral LD50 for rats is 1 - 2 mg per kilogram; the acute dermal LD50 for rats is 5 mg per kilogram; the hazards of vapour toxicity are high.

Use Effective against sap-sucking insects (aphids) and mites, but toxicity hazards are high.

Caution

1 This is a very poisonous substance (Part I p. 93) — full protective clothing must be worn.
2 Dangerous to fish, livestock, game, wild birds and mammals.
3 Pre-harvest interval for picking hops is 4 weeks.
4 Pre-access interval to treated areas is, for unprotected persons, 1 day; for livestock and poultry 4 weeks.

Formulation 'Terra Sytam' is a 50 per cent w.v. solution.

Dimethoate
(*00*-dimethyl *S*-methylcarbomoylmethyl phosphorodithioate)

Trade names 'Roger', 'Roxion', 'Perfekthion'

Properties A systemic and contact insecticide and acaricide, used mainly against fruit flies and aphids.

The pure compound is a white solid, only slightly soluble in water, soluble in most organic solvents except saturated hydrocarbons such as hexane. It is stable in aqueous solution and to sunlight, but is readily hydrolysed by aqueous alkali. It is incompatible with alkaline pesticides.

The acute oral LD50 for rats is 250 - 265 mg per kilogram.

Use Effective against aphids, psyllids, some flies, saw flies, woolly aphids and red spider mites.

Caution

1 Dangerous to bees.
2 Harmful to fish, livestock, game, wild birds and mammals.
3 Do not use on chrysanthemums, hops or on ornamental *Prunus* species.
4 Pre-harvest interval for edible crops is 1 week.
5 Pre-access interval for livestock to treated areas is 1 week.

Formulation 20 and 40 per cent e.c; 20 per cent w.p.; 5 per cent granules.

Disulfoton
(*00*-diethyl *S*-ethylthioethylphosphorodithioate)

Trade names 'Disyston', 'Dithio-systox'

Properties A systemic insecticide and acaricide used mainly as a seed dressing or as granules to protect seedlings from insect attack. It is metabolised in the plant to the sulphoxide and sulphone. It is a colourless oil with a characteristic odour, only slightly soluble in water, but readily soluble in most organic solvents. It is relatively stable to hydrolysis below pH 8·0.

The acute oral LD for male rats is 12·5 mg per kilogram.

Use Effective against aphids on vegetables and fruit; carrot fly; leafhoppers on rice, vegetables, cotton; some flies, leafminers and beetles.

Caution

1 This is a poisonous substance (Part II p. 93) — full protective clothing should be worn.
2 Dangerous to fish.
3 Pre-harvest interval for edible crops is 6 weeks.

Formulation 'Disyston' is 50 per cent impregnated on activated carbon; also 5 and 10 per cent granules based on Fuller's Earth (F.E.) or pumice (P.).

Ethion
(*000'0'*-tetraethyl *SS*-methylene di(phosphorodithioate))

Trade names 'Embathion', 'Hylemox'

Properties It is a non-systemic insecticide and

acaricide, used mainly in combination with petroleum oils on dormant fruit as an ovicide and scalecide. It is non-phytotoxic. It is a pale-coloured liquid very slightly soluble in most organic solvents including kerosene and petroleum oils. It is slowly oxidised in air and is subject to hydrolysis by both acids and alkalis.

The acute oral LD50 for rats is 208 mg per kilogram for the pure substance, and 96 mg per kilogram, technical grade.

Use Effective against eggs and dormant stages of pests on fruit trees. Some use against anthomyiid fly maggots on cereals and vegetables (in the soil).

Formulation 25 per cent w.p.; e.c.; 4 per cent dust; 50 per cent seed dressing.

Ethoate-methyl (*S*-ethylcarbamoylmethyl *00*-dimethyl phosphorodithioate)

Trade name 'Fitios'

Properties It is a systemic insecticide and acaricide with contact action, particularly effective against fruit flies. The pure compound is a white crystalline solid, almost insoluble in water, but soluble in benzene, trichloromethane, propanone and ethanol. It is stable in aqueous solution but hydrolysed by alkali.

The acute LD50 for male rats is 340 mg per kilogram: non-irritant.

Use Effective especially against olive fly, 60 g a.i. per litre, and fruit-flies, 0·6 g per litre, recommended for control of aphids, and red spider mites on fruit, arable and vegetable crops at rates of 0·1 – 4 g a.i. per 10 – 100 litres per hectare.

Caution

1 Harmful to fish, livestock, game, wild birds and mammals.
2 Dangerous to bees.
3 Pre-harvest interval for arable crops is 1 week.
4 Pre-access interval for animals to treated areas is 1 week.

Formulation 20 and 40 per cent e.c.; 25 per cent w.p.; 5 per cent dust and 5 per cent granules.

Fenitrothion (*00*-dimethyl 0-4 nitro-*m*-tolyl phosporothioate (I))

Trade names 'Accothion', 'Folithion', 'Sumithion'

Properties It is a contact and stomach insecticide, particularly effective against rice stem borers, and a selective acaricide but of low ovicidal activity. It is a brownish-yellow liquid, practically insoluble in water, but soluble in most organic solvents. It is hydrolysed by alkali and is of moderate persistence.

The acute oral LD for rats is 250 – 500 mg per kilogram.

Use Effective against lepidopterous larvae (rice stem borers especially), aphids, white capsids, psyllids; some fly larvae, some beetles, locusts, thrips and sawflies.

Caution

1 Harmful to fish, bees, livestock, game, wild birds and mammals.
2 Pre-harvest interval for edible crops is 2 – 3 weeks.
3 Pre-access interval for livestock to treated areas is 1 week.

Formulation 50 per cent e.c.; 40 and 50 per cent w.p.; 5, 3 and 2 per cent dusts.

Fenthion (*00*-dimethyl 0-4-methylthio-*m*-tolyl phosphorothioate (I))

Trade names 'Baytex', 'Lebaycid', 'Queletox'

Properties It is a contact and stomach insecticide with a useful penetrant action, which, by virtue of low volatility and stablility to hydrolysis, is of high persistence. It is a colourless liquid, practically insoluble in water, but readily soluble in most organic solvents. It is stable at temperatures up to 210 °C and is resistant to light and to alkaline hydrolysis.

The acute LD50 for male rats is 215 mg per kilogram. It is of greater toxicity to dogs and birds, and is used for the control of weaver birds in Africa.

Use Effective against fruit flies, many caterpillars, leafhoppers and plant bugs, aphids, thrips, mites, sawflies, some beetles and birds.

Caution

1 Harmful to birds and wildlife.

2 Pre-harvest interval for edible crops is 7 - 42 days according to country and crop.

Formulation 50, 40, 25 per cent w.p.; 60 per cent fogging concentrate; 50 per cent e.c.; 3 per cent dust; 'Queletox' for use against weaver birds.

Fonofos
(*O*-ethyl *S*-phenylethylphosphonodithioate)

Trade name 'Dyfonate'

Properties It is an insecticide particularly suitable for the control of soil maggots and other soil insects. Seed can be damaged by contact with fonofos. Persistence in soil is moderate: 8 weeks.

It is a pale-yellow liquid, practically insoluble in water, but miscible with most organic solvents such as kerosene, 1,2-dimethylbenzene, etc. It is stable under normal conditions.

Acute oral LD50 for male rats is 8 - 17 mg per kilogram.

Use Effective against most soil pests, such as root maggots, soil caterpillars, wireworms and other beetle larvae, crickets and symphylids.

Caution

1 This is a very poisonous substance (Part II p. 93) — full protective clothing should be worn.
2 Dangerous to fish.
3 Pre-harvest interval for edible crops is 6 weeks.

Formulation 5 and 10 per cent granules.

Formothion
(S-N-formyl-N-methylcarbamoylmethyl *OO*-dimethyl phosphorodithioate)

Trade names 'Anthio', 'Affix'

Properties It is a contact and systemic insecticide and acaricide effective against sap-sucking insects and mites. In plants, it is metabolised to dimethoate. In loamy soil the half-life is 14 days.

It is a yellow viscous oil or crystalline mass, slightly soluble in water, miscible with alcohols, trichloromethane, ethoxyethane, ketones and benzene. It is stable in non-polar solvents, but is hydrolysed by alkali and incompatible with alkaline pesticides.

The acute oral LD50 for male rats is 375 - 535 mg per kilogram.

Use Effective against many sap-sucking insects and mites, especially aphids and red spider mites, and some fly larvae.

Caution

1 Dangerous to bees.
2 Harmful to fish and livestock.
3 Pre-harvest interval for edible crops is 1 week.
4 Access of animals to treated areas is 1 week.

Formulation 25 per cent w/v e.c. is the usual formulation.

Heptenophos
(7-chlorobicyclo[3,2,0]-hepta-2,6-dien-6-yl dimethyl phosphate)

Trade name 'Hostaquick'

Properties A translocated insecticide with rapid initial action and short residual effect; it penetrates plant tissues and is quickly translocated in all directions. It is used against sap-sucking insects, some Diptera, and ectoparasites of domesticated animals.

A pale brown liquid, scarcely soluble in water but quite soluble in most organic solvents.

The acute oral LD50 for rats is 96 - 121 mg per kilogram.

Use Effective against many sap-sucking insects, but especially aphids.

Caution

1 This is a poisonous substance (Part III p. 93) — protective clothing should be worn.
2 Pre-harvest interval for edible crops is 1 day.
3 Dangerous to bees.
4 Harmful to fish.

Formulation For agricultural use only as an e.c. 500 g a.i. per litre.

Malathion
(*S*-1,2-di(ethoxycarbonyl)ethyl *OO*-dimethyl phosphorodithioate)

Trade names 'Malathon', Malathiozol', 'Cythion'

Properties It is a wide-spectrum non-systemic insecticide and acaricide, of brief to moderate per sistence and low mammalian toxicity. It is generally non-phytotoxic, but may damage cucurbits under glasshouse conditions, and various flower species.

It is a colourless or pale brown liquid, of slight solubility in water, miscible with most organic solvents though not in petroleum oils. Hydrolysis is rapid above pH 7·0 and below 5·0. It is incompatible with alkaline pesticides and corrosive to iron.

The acute oral LD50 for rats is 2 800 mg per kilogram.

Use Effective against aphids, thrips, leafhoppers, spider mites, mealybugs, scales, various beetles, caterpillars and flies.

Caution

1 Harmful to bees and fish.
2 To avoid possible taint to edible crops allow 4 days from application to harvest, (7 days for crops for processing).
3 Pre-harvest interval is 1 day.

Formulation As e.c. from 25 to 86 per cent (many of 60 per cent), 25 and 50 per cent w.p.; dusts of 4 per cent and as atomising concentrates (95 per cent) for ULV applications.

Mecarbam
(*S*-(*N*-ethoxycarbonyl-*N*-methylcarbamoylmethyl) *00*-diethyl phosphorodithioate)

Trade names 'Murfotox'. 'Pestan', 'Afos'

Properties An insecticide and acaricide with slight systemic properties used for control of Hemiptera and Diptera. At recommended rates it persists in soil for 4 – 6 weeks.

It is a pale brown oil, almost insoluble in water, slightly soluble in aliphatic hydrocarbons, miscible with alcohols, aromatic hydrocarbons, ketones and esters. It is easily hydrolysed. It is compatible with all but highly alkaline pesticides and slowly attacks metals.

The acute oral LD50 for rats is 36 mg per kilogram.

Use Effective against scale insects and other Hemiptera, olive fly and other fruit-flies, leaf hoppers and stem-flies of rice, and root-fly maggots on vegetables.

Formulation e.c. 68, 40 per cent; w.p. 25 per cent; dusts; 'Murfotox' oil 5 per cent in petroleum oil.

Menazon
(*S*-4,6-diamino-1,3,5-triazin-2-ylmethyl *00*-dimethyl phosphorodithioate)

Trade names 'Saphizon', 'Saphiol', 'Sayfos'

Properties It is a systemic insecticide, used mainly against aphids. It is regarded as non-phytotoxic.

It forms colourless insoluble crystals, stable up to 35 °C. It is weakly basic and is compatible with all but strongly alkaline pesticides. It may be decomposed by the reactive surfaces of some 'inert' fillers.

The acute oral LD50 for female rats is 1 950 mg per kilogram.

Use Most frequently used against aphids, as a seed dressing, also as a drench and a root dip. It is used against woolly aphid on apple.

Caution

1 Harmful to bees and livestock.
2 Pre-harvest interval for edible crops is 3 weeks.

Formulation Seed dressings as 50 and 80 per cent; 70 per cent w.p.

Methidathion
(*S*-2,3-dihydro-5-methoxy-2-oxo-1,3,4-thiadiazol-3-ylmethyl *00*-dimethyl phosphorodithioate)

Trade name 'Supracide'

Properties It is a non-systemic insecticide, with some acaricidal activity. It is capable of foliar penetration. Non-phytotoxic to all plants tested, it is rapidly metabolised and excreted by plants and animals.

It forms colourless crystals, of slight solubility in water, readily soluble in propanone, benzene and methanol. It is stable in neutral and weakly acid media, but much less stable in alkali. It is compatible with many fungicides and acaricides.

Acute oral LD50 for rats is 25 – 48 mg per kilogram.

Use Used against lepidopterous larvae; foliar penetration enables it to be used against leaf rollers. Used on a wide variety of crops against leaf eating and sucking insects and mites, especially against scale insects.

Caution

1 This is a poisonous substance (Part II p. 93) — protective clothing should be worn.
2 Harmful to bees and livestock.
3 Dangerous to fish.
4 Pre-harvest interval for edible crops is 3 weeks.
5 Pre-access interval for livestock to treated areas is 2 weeks.

Formulation 20 and 40 per cent e.c. and w.p.

Mevinphos
(2-methoxy-carbonyl-1-methylvinyl dimethyl phosphate)

Trade name 'Phosdrin'

Properties It is a contact and systemic insecticide and acaricide of short persistence. Although non-persistent, its high initial kill provides a relatively long period before the pest builds up again. It is non-phytotoxic.

The technical product is a pale yellow liquid, miscible with water, alcohols, ketones, chlorinated hydrocarbons, aromatic hydrocarbons, but only slightly soluble in aliphatic hydrocarbons. Stable at ordinary temperatures, but hydrolysed in aqueous solution, and rapidly decomposed by alkalis. It is thus incompatible with alkaline fertilisers and pesticides. It is corrosive to cast iron, mild and some stainless steels, and brass; relatively non-corrosive to copper, nickel and aluminium; non-corrosive to glass, and many plastics, but passes slowly through thin films of polythene.

Acute oral LD50 for rats is 3·7 - 12 mg per kilogram.

Use Effective against sap-feeding insects (aphids, etc.) at 20 - 45 g per hectare, mites and beetles at 30 - 60 g per hectare, caterpillars at 45 - 90 g per hectare and some fly larvae. Especially useful for giving rapid kill close to harvest.

Caution

1 This is a poisonous substance (Part II p.93) — full protective clothing should be worn.
2 Dangerous to livestock, bees, fish, and game.
3 Pre-harvest interval to edible crops is 3 days.
4 Pre-access interval to treated areas is 1 day.

Formulation It is water soluble and formulation is unnecessary, but e.c. of 5, 10, 18, 24, 58 and 50 per cent technical are available, together with dusts and w.p.s.

Monocrotophos
(dimethyl 1-methyl-2-methylcarbamoylvinyl phosphate)

Trade names 'Nuvacron', 'Azodrin'

Properties A fast-acting insecticide with both systemic and contact action. It is used against a wide range of pests on a variety of crops and has persistence of 1 - 2 weeks. It is phytotoxic under cool conditions to some apples, cherries and sorghum varieties and is incompatible with alkaline pesticides.

It is a crystalline solid, miscible with water, soluble in propanone and ethanol, sparingly soluble in 1,2-dimethylbenzene but almost insoluble in kerosene and diesel oils. Unstable in low molecular weight alcohols and glycols. Corrosive to iron, steel and brass, but does not attack glass, aluminium and stainless steel.

Acute LD50 for rats is 13·23 mg per kilogram.

Use Effective against a wide range of pests including mites, bugs, leaf-eating beetles, leafminers and caterpillars.

Caution Dangerous to fish and livestock.

Formulation Water-miscible concentrates contain 200 - 600 g a.i. per litre.

Naled
(1,2-dibromo-2,2-dichloroethyl dimethyl phosphate)

Trade name 'Dibrom'

Properties A non-systemic contact and stomach insecticide and acaricide with some fumigant action. It is used mainly under glass and in mushroom houses.

The technical product is a yellow liquid, insoluble in water, slightly soluble in aliphatic solvents, and readily soluble in aromatic solvents. It is stable under anhydrous conditions, but rapidly hydrolysed in water (90 - 100 per cent in 48 hours at room temperature), and by alkali; stable in glass containers, but in the presence of metals and reducing agents, rapidly loses bromine and reverts to dichlorvos.

Acute LD50 for rats is 430 mg per kilogram.

Use For glasshouse and mushroom pests.

Formulation 4 per cent dust and e.c. 0·8 per litre.

Omethoate
(*00*-dimethyl *S*-methylcarbamoylmethyl phosphorothioate)

Trade name 'Folimat'

Properties A systemic insecticide and acaricide with a broad range of action and little phytotoxicity, except to some peach varieties.

It is a colourless liquid, readily soluble in water, propanone, ethanol and many hydrocarbons; insoluble in light petroleum; hydrolysed by alkali.

Acute oral LD50 for rats is about 50 mg per kilogram.

Use Effective against a wide range of insects, particularly caterpillars, and Homoptera, Orthoptera, thrips, some beetles and phytophagous mites.

Caution

1 Harmful to bees.
2 Damaging to some varieties of peach.
3 Pre-harvest interval on edible crops is 21-28 days.

Formulations Include e.c. and granules with a range of a.i. contents.

Oxydemeton-methyl
(S-2-ethylsulphinylethyl 00-dimethyl phosphorothioate)

Trade name 'Metasystox-R'

Properties A systemic and contact insecticide and acaricide, used against sap-sucking insects mites, with a fast kill and moderate persistence.

It is a clear brown liquid, miscible with water and soluble in most organic solvents except light petroleum. It is hydrolysed by alkali.

The acute oral LD50 for male rats is 65 mg per kilogram.

Use Effective against aphids, and red spider mites on most crops, also leafhoppers, psyllids, whiteflies, thrips, some flies and sawflies. Only a limited effect on *Brassica* aphids.

Caution

1 This is a poisonous substance (Part III p. 93) — and protective clothing should be worn.
2 Harmful to bees, livestock, fish, game and wild animals.
3 Pre-harvest interval for edible crops is 2-3 weeks.
4 Pre-access interval for livestock to treated areas is 2 weeks.

Formulation 25 and 50 per cent e.c. of various a.i. contents.

Oxydisulfoton
(*00*-diethyl *S*-2-ethylsulphinylethyl phosphorodithioate)

Trade name 'Disyston-S'

Properties A systemic insecticide and acaricide particularly suitable for seed treatment against virus vectors.

It is a pale-coloured liquid, slightly soluble in water, readily soluble in most organic solvents.

The acute oral LD50 for rats is about 3·5 mg per kilogram.

Use Particularly effective against sap-sucking insects and mites.

Caution This is very poisonous substance (Part II p. 93) — full protective clothing should be worn.

Formulation As seed dressings of various a.i. contents, also as e.c. and granules.

Parathion
(*00*-diethyl *0*-4-nitrophenyl phosphorothioate)

Trade names 'Folidol', 'Bladan', 'Thiophos', 'Fosferno'

Properties A non-systemic, contact and stomach insecticide and acaricide with some fumigant action. It is non-phytotoxic except to some ornamentals, and under certain weather conditions, to pears and some apple varieties.

It is a pale yellow liquid, scarcely soluble in water, slightly soluble in petroleum oil, but miscible with most organic solvents. In alkaline solution it rapidly hydrolyses; on heating it isomerises.

The acute oral LD50 for male rats is 13 mg per kilogram; for females 3·6 mg per kilogram.

Use Effective against most Homoptera, Diptera, springtails, mites, millipedes and some nematodes. Prolonged use may result in extensive destruction of predators and parasites.

Caution

1 This is a very poisonous substance (Part II p. 93) — full protective clothing should be worn.

2 Dangerous to bees, fish, livestock, game, wild birds and mammals.
3 Pre-harvest interval for edible crops is 4 weeks.
4 Pre-access interval for livestock to treated areas is 10 days.

Formulation To w.p. and e.c. of various a.i. contents; also to dusts, smokes and aerosols.

Parathion-methyl
(*OO*-dimethyl *O*-4-nitrophenyl phosphorothioate (I))

Trade names 'Dalf', 'Metacide', 'Folidol-M'

Properties A non-systemic, contact and stomach insecticide, with some fumigant action. Its range of action is similar to that of parathion but of lower mammalian toxicity. It is non-phytotoxic.

The technical product is a brown liquid of about 80 per cent purity, scarcely soluble in water, slightly soluble in light petroleum and mineral oils, but soluble in most other organic solvents. It is hydrolysed by alkali at a faster rate than parathion, and readily isomerises on heating. Compatible with most other pesticides.

The acute oral LD50 for male rats is 14 mg per kilogram and for female rats, 24 mg per kilogram. It is hazardous to wildlife but is of brief persistence.

Caution
1 This is a poisonous substance, so protective clothing should be worn.
2 Hazardous to wild life.

Use As for parathion but not effective against Acarina.

Formulation To e.c. and dusts of various a.i. contents; 'Nitrox 80' is an 80 per cent solution in an aromatic petroleum solvent.

Phenisobromolate
See bromopropylate (p. 115).

Phenthoate
(*S*-α-ethoxycarbonylbenzyl *OO*-dimethyl phosphorodithioate)

Trade names 'Elsan', 'Cidial'

Properties A non-systemic insecticide and acaricide with contact and stomach action. It may be phytotoxic to some peach, fig and grape and varieties, and may discolour some red-skinned apple varieties.

It is a crystalline solid, almost insoluble in water, but miscible with most organic solvents.

The acute oral LD50 for rats is 250 – 300 mg per kilogram.

Use Effective against caterpillars, aphids, jassids, mites, and is also used for the protection of stored grain.

Formulation 50 per cent technical, 5 per cent in mineral oil; 40 per cent w.p.; 2 per cent granules; 85 per cent pure compound.

Phorate
(*OO*-diethyl *S*-ethylthiomethyl phosphorodithioate)

Trade name 'Thimet'

Properties A persistent, systemic insecticide used on granular and e.c. formulations for the protection of seedlings from sap-feeding and soil insects; some fumigant action.

It is a clear liquid, only slightly soluble in water, but miscible with tetrachloromethane, dioxane, 1,2-dimethylbenzene and vegetable oils. It is hydrolysed by alkalis and in the presence of moisture.

The acute oral LD50 for male rats is 3·7 mg per kilogram; for females 1·6 mg per kilogram.

Use Effective against aphids, wireworms, various fly maggots (fruit, carrot), capsids, leafhoppers, various weevils.

Caution
1 This is a very poisonous substance (Part II, p. 93) — full protective clothing should be worn.
2 Dangerous to fish and livestock.
3 Pre-harvest interval for edible crops is 6 weeks.
4 Pre-access interval for livestock to treated areas is 6 weeks.

Formulation As e.c. of various a.i. contents; 5, 10 and 15 per cent granules.

Phosalone
(*S*-6-chlor-2-oxobenzoxazolin-3-ylmethyl *OO*-diethyl phosphorodithioate)

Trade name 'Zolone'

Properties A non-systemic insecticide and

acaricide used on deciduous tree fruits, field and market garden crops, against a wide range of pests. It persists on plants for about 2 weeks before being hydrolysed.

It forms colourless crystals, insoluble in water and light petroleum, but soluble in propanone, benzene, trichloromethane, ethanol, methanol, methylbenzene and 1,2-dimethylbenzene. It is stable under normal storage conditions, non-corrosive, and compatible with most other pesticides.

The acute oral LD50 for male rats is 150 mg per kilogram.

Use Effective against a wide spectrum of pests including caterpillars on fruit, cotton bollworms, fruit-fly maggots, aphids, psyllids, jassids, thrips, various weevils and red spider mites.

Caution

1 Harmful to bees, fish and livestock.
2 Pre-harvest interval for edible crops is 3 weeks.
3 Pre-access interval for livestock is 4 weeks.

Formulation 30, 33, 35 per cent e.c.; 30 per cent w.p.; 2·5 and 4 per cent dusts. Various e.c. formulations under heading 'Zolone DT' with DDT for use on cotton.

Phosmet
(*00*-dimethyl *S*-phthalimidomethyl phosphorodithioate)

Trade name 'Imidan'

Properties A non-systemic acaricide and insecticide, used at concentrations safe for a variety of predators of mites and thus useful for integrated control programmes.

It is a while crystalline solid, with an offensive odour, scarcely soluble in water, but more than 10 per cent soluble in propanone, dichloromethane and 1,2-dimethylbenzene.

The acute oral LD50 for male rats is 230 mg per kilogram; it is readily degraded both in laboratory animals and in the environment.

Use Used mainly against phytophagous mites; when used as recommended it should not affect mite predators.

Formulation 20 and 30 per cent e.c. and 50 per cent w.p. Storage above 45 °C may lead to decomposition.

Phosphamidon
(2-chloro-2-diethylcarbamoyl-1-methylvinyl dimethyl phosphate)

Trade names 'Dimecron', 'Dicron'

Properties A systemic insecticide and acaricide, rapidly absorbed by the plant, but only a little contact action. Non-tainting, and non-phytotoxic except to some cherry varieties and sorghum varieties related to red swazi.

It is a pale yellow oil, miscible with water, and readily soluble in most organic solvents except saturated hydrocarbons. It is stable in neutral and acid media but is hydrolysed by alkali. Compatible with all but highly alkaline pesticides. It corrodes iron, tin-plate and aluminium, and is packed in polythene containers.

The acute oral LD50 for rats is 28·3 mg per kilogram. The half-life in plant is about 2 days.

Use Effective against sap-feeding and leaf-eating insects, particularly aphids, many caterpillars (but not Noctuidae), rice stem-borers, thrips, Colorado beetle and other beetles, grasshoppers, sawflies, fly larvae and phytophagous mites.

Caution

1 This is a poisonous substance (Part III p. 93) — protective clothing should be worn.
2 Dangerous to bees.
3 Harmful to livestock.
4 Pre-harvest interval for edible crops is 3 weeks.
5 Pre-access interval for livestock to treated areas is 2 weeks.

Formulation 'Dimecron 20': 20 kg per 100 litres in propan-2-o1, the water content rigidly controlled to delay hydrolysis; similarly for 'Dimecron 50' and 100; 50 per cent w.p.

Phoxim
(00-diethyl α-cyanobenzylideneamino-oxyphosphonothioate

Trade names 'Baythion', 'Volaton'

Properties An insecticide of brief persistence and no systemic action with low mammalian toxicity, effective against a broad range of insects.

It is a yellow liquid virtually insoluble in water, slightly soluble in light petroleum, soluble in alcohols, ketones, and aromatic hydrocarbons.

Stable in water and acid media, but unstable in alkali. Believed to be compatible with most non-alkaline pesticides.

The acute oral LD50 for rats is more than 2 000 mg per kilogram.

Use Effective against a broad range of insects, especially stored-product pests and insects affecting man. Also successfully used against soil pests (dipterous maggots, rootworms and wireworms); ULV applications are effective against grasshoppers.

Caution

1 Dangerous to bees by contact and vapour effect.
2 Harmful to fish.

Formulation 50 per cent e.c; 5 per cent granules; and a concentrate for ULV application. Other experimental formulations are under test. 'Baythion' is used against pests of man and stored products and 'Volaton' is for agricultural use.

Pirimiphos-ethyl
(*O*-2-diethylamino-6-methylpyrimidin-4-yl *OO*-diethyl phosphorothioate)

Trade name 'Pirimicid'

Properties It is a broad-spectrum insecticide, particularly against soil-inhabiting Diptera and Coleoptera. No phytotoxicity has been recorded using recommended rates, although high rates of seed dressings have resulted in seedling abnormalities. Stable for 5 days at 80 °C.

In the pure state it is a pale straw-coloured liquid; insoluble in water but miscible with most organic solvents. It is corrosive to iron and unprotected tin-plate.

Acute oral LD50 for rats is 140 - 200 mg per kilogram.

Use Effective against soil-inhabiting Diptera and Coleoptera; effective as foliage spray at conventional rates against species of Lepidoptera, Coleoptera, Homoptera and Tetranychidae.

Caution

1 Dangerous to bees.
2 Hazardous to birds and wild animals.

Formulation A 20 per cent s.d.; encapsulated, 200 g a.i. per litre, 500 g per litre e.c.; granules 50 and 100 g per kilogram with 5 and 10 per cent thiram added.

Pirimiphos-methyl
(*O*-2-diethylamino-6-methylpyrimidin-4-yl *OO*-dimethyl phosphorothioate)

Trade names 'Actellic', 'Blax'

Properties It is a fast-acting, broad-spectrum insecticide of limited persistence, with both contact and fumigant action. It is non-phytotoxic. It penetrates leaf tissue to the extent that an insect on one side of the leaf is killed by chemical applied to the other side; there is also a slight systemic action.

It is a straw-coloured liquid, insoluble in water, but soluble in most organic solvents. It is decomposed by strong acids and alkalis, does not corrode brass, stainless steel, nylon or aluminium.

Acute oral LD50 for female rats is about 800 mg per kilogram (low mammalian toxicity).

Use Effective against species of Lepidoptera, Coleoptera, aphids, Tetranychidae, and many other crop pests (Homoptera, Heteroptera, Diptera, Thysanoptera, Orthoptera).

Caution Low mammalian toxicity means that restrictions for use are few.

Formulation 25 and 50 per cent e.c.; 5 and 10 per cent granules, and 1 000 g per litre and 500 g per litre ULV formulations.

Profenofos
(*O*-(4-bromo-2-chlorophenyl)*O*-ethyl *S*-propyl phosphorothioate)

Trade name 'CURACRON'

Properties A broad-spectrum insecticide; non-systemic; with contact and stomach action; used against many crop pests.

It is a pale yellow liquid, barely soluble in water, but miscible with most organic solvents; fairly stable under neutral and slightly acid conditions.

The acute oral LD50 for rats is 358 mg per kilogram.

Use Effective against many cotton and vegetable pests, both insects and mites; rates used are 250 - 500 g a.i. per hectare for sucking insects and mites, and 400 - 1 200 g a.i. per hectare for chewing insects.

Formulation e.c. 500 g a.i. per litre, 400 g a.i. per litre; ULV 250 g a.i. per litre; granules 50 g a.i. per kilogram; and mixtures with chlordimeform.

Prothoate
(*OO*-diethyl *S*-isopropylcarbamoylmethyl phosphorodithioate)

Trade name 'Fac'

Properties An acaricide and insecticide with systemic action, used mainly against phytophagous mites and some sap-sucking insects.

It is an almost colourless crystalline solid, virtually insoluble in water, but miscible with most organic solvents. It is stable in neutral, moderately acid and slightly alkaline media, but is rapidly decomposed in strong alkali.

The acute oral LD50 for male rats is 8 mg per kilogram.

Use Effective for protection of fruit, citrus and vegetable crops from tetranychid and some eriophyid mites, and some insects, notably aphids, Tingidae, Psyllidae and Thysanoptera.

Caution

1 This is a very poisonous substance, and full protective clothing should be worn.

2 Dangerous to fish, livestock and game.

Formulation 3, 20 and 40 per cent technical; 5 per cent granules.

Quinomethionate
(6-methyl-2-oxo-1,3-dithiolo-(6-methyl-2-oxo-1,3-dithiolo-[4,5-*b*]quinoxaline))

Trade name 'Morestan'

Properties A selective, non-systemic acaricide and a fungicide specific to powdery mildews.

It forms yellow crystals, practically insoluble in water and sparingly soluble in organic solvents. In chemical properties it is closely related to thioquinox, but is more stable to oxidation.

The acute oral LD50 for rats is 2 500 - 3 000 mg per kilogram.

Use Controls red spider and other phytophagous mites on a wide range of crops, and powdery mildews.

Caution

1 Harmless to bees.

2 Blackcurrant varieties may be damaged by sprays.

3 Pre-harvest interval for edible crops is 3-28 days, according to crops and country.

Formulation 25 per cent w.p.; and also formulated as smokes.

Schradan
(di(tetramethylphosphorodiamidic) anhydride)

Trade names 'Sytam', 'Pestox 3'

Properties A systemic insecticide and acaricide with little contact effect; effective against sap-feeding insects and mites; non-phytotoxic at insecticidal concentrations.

It is a brown viscous liquid, miscible with water and most organic solvents; slightly soluble in petroleum oils, and readily extracted from aqueous solution by trichloromethane. Stable to water and alkali, but hydrolysed under acid conditions.

The acute oral LD50 for male rats is 9·1 mg per kilogram.

Use Used mainly against aphids and red spider mites on a variety of crops.

Caution

1 This is a very poisonous substance (Part II p. 93) — full protective clothing should be worn.

2 Dangerous to fish, livestock, game, wild birds and mammals.

3 Safe to bees.

4 Pre-harvest interval for edible crops is 4-6 weeks.

5 Pre-access interval for livestock to treated areas is 4 weeks.

Formulation 30 per cent aqueous solution; also anhydrous, 75-80 or 60 per cent with anhydrous surfactant.

TEPP
(tetraethyl pyrophosphate (I))

Trade names 'Nifos T', 'Vapotone'

Properties A non-systemic aphicide and acaricide of brief persistence; very high mammalian toxicity.

It is a colourless, hygroscopic liquid, miscible with water and most organic solvents, but only slightly soluble in petroleum oils. It is rapidly hydrolysed by water and is corrosive to most metals. It is metabolised very fast by animals.

The acute oral LD50 for rats is 1·62 mg per kilogram; the acute dermal LD50 is 2·4 mg per kilogram.

Use Used only against aphids and phytophagous mites.

Caution

1 An extremely poisonous pesticide — full protective clothing must be worn. It has high mammalian dermal toxicity.
2 Dangerous to livestock, game, wild birds and mammals.

Formulation For agricultural purposes TEPP refers to a mixture of polyphosphates containing at least 40 per cent tetraethyl pyrophosphate. It is used as an aerosol solution in methyl chloride.

Terbufos
(*S-tert*-butylthiomethyl *OO*-diethyl phosphorodithioate)

Trade name 'Counter'

Properties An insecticide with strong inital and residual activity against soil insects.

It is a pale yellow liquid, almost insoluble in water, but soluble in many organic solvents. It decomposes on prolonged heating at high temperatures, at low pH and with strong alkalis.

The acute oral LD50 for rats is 1·6 – 4·5 mg per kilogram; the acute dermal LD50 for rats is from 1·0 – 7·4 mg per kilogram.

Use Effective against soil insects of different types.

Caution

1 This is a very poisonous substance, and full protective clothing must be worn.
2 Can be absorbed through the skin.
3 Dangerous to wildlife and game.

Formulation Only as granules, from 20 – 150 g a.i. per kilogram.

Tetrachlorvinphos
((*z*)-2-chloro-1-(2,4,5,-trichlorophenyl)vinyl dimethyl phosphate)

Trade name 'Gardona'

Properties A selective insecticide used against lepidopterous and dipterous pests on the aerial parts of crops; very low mammalian toxicity.

It is a white crystalline solid, scarcely soluble in water, but soluble in chloroform (40 per cent w/w), methyl chloride (40 per cent); less so in 1,2-dimethylbenzene (15 per cent) and propanone (20 per cent). It is heat stable, but slowly hydrolysed by water, particularly under alkaline conditions.

The acute oral LD50 for rats is 4 – 5 mg per kilogram.

Use Effective against lepidopterous and dipterous pests of fruit, rice, vegetables, cotton and maize. With certain exceptions it does not show high activity against Hemiptera, and because of rapid breakdown is not effective in the soil. Shows promise against pests of stored food products.

Formulation 240 g per litre e.c.; 50 per cent and 70 per cent w.p.; and 5 per cent granules.

Thiometon
(*S*-2-ethylthioethyl *OO*-dimethyl phosphorodithioate)

Trade name 'Ekatin'

Properties A systemic insecticide and acaricide suitable for control of sucking insects and mites in orchards, vineyards, and on beet. At 0·02 a.i. per cent systemic effects persist for 2 – 3 weeks.

It is a colourless oil, fairly soluble in water, slightly soluble in light petroleum oils but soluble in most other organic solvents.

The acute oral LD50 for rats is 120 – 130 mg per kilogram.

Use Mainly used for aphid control (and also mites) on fruit, potatoes, beet, cereals and vegetables.

Caution

1 This is a poisonous substance (Part III p. 93) — protective clothing should be worn.
2 Harmful to bees, fish, livestock, game, wild birds and mammals.
3 Pre-harvest interval for edible crops is 3 weeks.
4 Pre-access interval for livestock to treated areas is 2 weeks.

Formulation 20 per cent e.c.; coloured blue; and a dry spray dust.

Thionazin
(*OO*-diethyl *O*-pyrazin-2-yl phosphorothioate)

Trade names 'Nemafos', 'Zinophos'

Properties A soil insecticide and nematicide, of relatively brief persistence.

It is a brown liquid, fairly soluble in water, miscible with most organic solvents; readily hydrolysed by alkali.

The acute oral LD50 for rats is 12 mg per kilogram and the acute dermal LD50 is 11 mg per kilogram.

Use Soil application is effective against symphylids and cabbage root fly (drench); granules recommended for use on cotton, cucurbits, groundnuts, *Brassica* spp. and tomato.

Caution

1 This is a very poisonous substance (Part II p. 93) — full protective clothing should be worn.
2 Dangerous to fish and livestock.
3 Pre-access interval for livestock to treated areas is 8 weeks.

Formulation 25 and 46 per cent e.c.; 5 and 10 per cent granules.

Thioquinox
(2-thio-1,3-dithiolo[4,5-*b*]quinoxaline)

Trade names 'Eradex'

Properties A non-systemic acaricide effective against eggs; and a fungicide specific against powdery mildews.

It is a brown powder, practically insoluble in water and most organic solvents, but is slightly soluble in ethanol and propanone. Stable to heat and light; resistant to hydrolysis, but susceptible to oxidation without any reduction of biological activity.

The acute oral LD50 for rats is 3 400 mg per kilogram.

Use Specific against eggs of phytophagous mites; and the powdery mildews.

Caution Some dermal irritation may be caused to some operators.

Formulation 50 per cent w.p.

Triazophos
(*OO*-diethyl *O*-1-phenyl-1,2,4-triazol-3-yl phosphorothioate)

Trade name 'Hostathion'

Properties A broad-spectrum insecticide and acaricide with some nematicidal properties; either foliar or soil application. It can penetrate plant tissues but has no systemic action.

A pale brown liquid, with little solubility in water, but soluble in most organic solvents.

The acute oral LD50 for rats is 82 mg per kilogram.

Use As a foliar spray it controls aphids on fruit (at 75 - 125 g a.i. per 100 litres) and on cereals (at 320 - 600 g a.i. (40 per cent e.c.) per hectare). Incorporated into the soil it has been used to control wireworms and some cutworms at 1 - 2 kg a.i. per hectare.

Caution

1 This is a poisonous substance, so full protective clothing should be worn.
2 Dangerous to bees, wildlife and game.

Formulation To e.c. 400 g a.i. per litre; ULV concentrates 250 and 400 g a.i. per litre; w.p. 300 g a.i. per kilogram; granules 20 and 50 g a.i. per kilogram.

Trichloronate
(*O*-2,4,5-trichlorophenylethylphosphonothioate)

Trade names 'Agritox', 'Agrisil', 'Phytosol'

Properties A non-systemic insecticide persistent in soil recommended for control of root maggots, wireworms and other soil insects. It acts as a stomach and contact poison. Recently shown to be effective against stem eelworm. Foliar uses are under test. Soil applications have given up to 5 months residual control.

A brown liquid, practically insoluble in water, but soluble in propanone, ethanol, aromatic solvents, kerosene, and chlorinated hydrocarbons. It is hydrolysed by alkali.

The acute oral LD50 for rats is 16 - 37 mg per kilogram. The acute dermal LD50 for rats is 0·25 ml per kilogram.

Use Effective against root maggots, wireworms and other soil and grassland insects. Also gives control of some nematodes. Very persistent in soil. Probably effective against cutworms, termites and *Collembola* spp.

Caution

1 A very poisonous substance, so full protective clothing must be worn. Very easily absorbed through mammalian skin.
2 Use in some countries prohibited because of high toxicity and excessive persistence.
3 Pre-harvest interval for edible crops varies from 4 weeks (Norway) to 8 weeks (Holland).
4 Dangerous to fish, livestock, game, wild birds and mammals.

Formulation 50 per cent e.c. and granules of various a.i. content (2·5, 7·5 per cent), also as a 20 per cent seed dressing.

Trichlorphon
(dimethyl 2,2,2-trichlor-1-hydroxyethylphosphonate)

Trade names 'Dipterex', 'Negavon', 'Tugon'

Properties A contact and stomach insecticide with penetrant action, recommended for use against flies, some bugs, some beetles, lepidopterous larvae, and ectoparasites of domestic animals. It is readily metabolised to dichlorvos and this is thought to account for its activity.

It is a white crystalline powder; quite soluble in water, insoluble in petroleum oils, poorly soluble in tetrachloromethane and ethoxyethane but soluble in benzene, ethanol, and most chlorinated hydrocarbons. It is stable at room temperature but is decomposed by water at higher temperatures (and in acid media) to form dichlorvos.

The acute oral LD50 for male rats is 630 mg per kilogram.

Use Effective against most lepidopterous larvae, flies and fly maggots, some Homoptera, many Heteroptera and some beetles. Used for household and veterinary pests, under different trade names.

Vamidothion
0,0-dimethyl-5,2-(1-methylcarbamoylethylthio)-ethyl-1-phosphorothioate

Trade names 'Kilval', 'Kilvar', 'Trucidor'

Properties Off-white waxy solid, very soluble in most organic solvents, practically insoluble in light petroleum, and cyclohexane; solubility in water is 4 kg per litre. Compatible with most pesticides and is non-corrosive. Undergoes slight decomposition at room temperature.

Acute oral LD50 for rats is 64–105 mg per kilogram.

Use Systemic insecticide and acaricide of high persistence against *Eriosoma lanigerum*, used for control of sap-feeding insects and mites on apples, pears, peaches, plums, hops, rice, cotton, etc. at 37–50 g a.i. per 100 litres.

Caution Goldfish survived 14 days in water containing 10 g Vamidothion per litre.

Formulation 400 g a.i. per litre.

Carbamates

The successful development of the organophosphates as pesticides has directed attention to other compounds which act as an anticholinesterases. This led to the discovery of carbaryl with its broad spectrum of activity. It appears that the action of the carbamates on both insects and mammals is an inhibition of the enzyme cholinesterase.

Bufencarb
(3-(1-methylbutyl)phenylmethylcarbamate (I))

Trade name 'BUX'

Properties A non-persistent carbamate insecticide, effective against a range of soil and foliage insects. Soil degradation is fairly rapid, with no seasonal accumulation expected following use of granules.

The technical product is a yellow solid, of low melting point, almost insoluble in water but very soluble in methanol and 1,2-dimethylbenzene, but less so in aliphatic hydrocarbons such as hexane. Stable in neutral or acid solutions, but the rate of hydrolysis increases with rise in either temperature or pH.

The acute oral LD50 for rats is 87 mg per kilogram; acute dermal LD50 for rabbits is 680 mg per kilogram.

Use Active at rates of 0·5–2 kg a.i. per hectare, against a range of soil insects, such as corn rootworm, rice water weevil, root mealybugs; and foliage insects, such as rice leafhoppers and planthoppers; and against stem-borers.

Caution

1 This is a very poisonous substance, and full protective clothing should be worn.
2 It can be absorbed through the skin.
3 Dangerous to fish.

Formulation To e.c 240 g a.i. per litre and 360 g a.i. per litre; dusts 20 and 40 g a.i. per kilogram; and granules 100 g a.i. per kilogram.

Carbaryl
(1-naphthyl methylcarbamate (I))

Trade names 'Sevin', 'Carbaryl 85'

Properties A contact insecticide with slight systemic properties and broad-spectrum activity. No evidence of any phytotoxicity at recommended rates. Has growth regulatory properties and may be used for fruit thinning (apples). Also used for killing earthworms in turf. Very persistent.

It is a white crystalline solid, barely soluble in water, but soluble in most polar organic solvent (such as dimethyl sulphoxide). It is stable to light, heat and hydrolysis. Compatible with most pesticides, except those strongly alkaline such as lime-sulphur or Bordeaux mixture, which cause hydrolysis.

The acute oral LD50 for male rats is 850 mg per kilogram.

Use Effective against many insect pests, especially caterpillars, midges, beetles, Orthoptera, capsids and other bugs. Generally more effective against chewing insects than sap suckers.

Caution

1 Dangerous to bees.
2 Harmful to fish.
3 Pre-harvest interval for edible crops is 1 week.

Formulation To 50 and 85 per cent w.p.; 5 and 10 per cent dusts.

Carbofuran
(2,3-dihydro-2,2-dimethylbenzofuran-7-yl-methylcarbamate)

Trade names 'Furadan', 'CURATERR'

Properties A broad-spectrum systemic insecticide, acaricide and nematicide, used against both foliage feeding insects and mites and soil pests. In plants its half-life is less than 5 days; in soil the half-life is 30 - 60 days.

It is a white, odourless, crystalline solid, fairly soluble in water and propanone; non-inflammable, but unstable in alkaline media.

The acute oral LD50 for rats is 8 - 14 mg per kilogram.

Use Effective against various sap-feeding insects and mites, especially leafhoppers on rice, root-eating caterpillars and beetle larvae on cereals.

Caution

1 This is a poisonous substance, and protective clothing must be worn.
2 Harmful to fish.

Formulation 75 per cent w.p.; 4 per cent flowable paste; 2,3 and 10 per cent granules.

Methiocarb
(4-methylthio-3,5-xylyl methylcarbamate (I))

Trade names 'Mesurol', 'Draza'

Properties A non-systemic insecticide and acaricide, with a broad spectrum of activity. It is persistent and acts as a powerful molluscicide.

It is a white crystalline powder, practically insoluble in water, but soluble in most organic solvents.Hydrolysed by alkali.

The acute oral LD50 for male rats is 100 mg per kilogram.

Use Mainly used for control of slugs and snails, can control cutworms and various beetles.

Caution

1 Harmful to fish.
2 Pre-harvest interval for edible crops is 7 days.
3 Pre-access interval for livestock to treated areas is 7 days.

Formulation To 50 and 70 per cent w.p; 'Draza', 4 per cent granules.

Methomyl
(*S*-methyl *N*-(methylcarbamoyloxy) thioacetamidate)

Trade name 'Lannate'

Properties A broad-spectrum, systemic and contact insecticide and acaricide. It is also effective as a nematicide.

It is a white crystalline solid, slightly soluble in water, soluble in propanone, ethanol, methanol; the aqueous solution is non-corrosive. It is stable in solid form and aqueous solution under normal conditions, but decomposes in moist soil.

The acute oral LD50 for male rats is 130 mg kilogram of the 25 per cent w.p. Not a skin irritant.

Use As a soil treatment it has given systemic control of certain insects and nematodes. It is of promise as a foliar spray for many insects such as aphids, Colorado beetle, leafrollers and many other caterpillars, and red spider mites.

Caution

1 This is a poisonous substance, and protective clothing should be worn.
2 The pre-harvest interval is about 3 weeks.

Formulation 25 and 90 per cent w.p.

Oxamyl
(*NN*-dimethyl-α-methylcarbamoyloxyimino-α-(methylthio)acetamide (I))

Trade name 'Vydate'

Properties A carbamate insecticide and nematicide with both contact and systemic action; moderate residual effect. Will control nematodes from both soil and foliar application.

A white crystalline solid, stable in solid form and most solutions, but decomposes to non-toxic materials in natural waters and soil. Fairly soluble in water, but more soluble in ethanol, propanone and methanol. Aqueous solution is non-corrosive. Aeration, sunlight, alkalinity and higher temperature increase rate of decomposition.

The acute oral LD50 for male rats is 5·4 mg per kilogram.

Use Effective against thrips, aphids, flea beetles, leaf beetles, leaf miners, and mites as foliar sprays at rates of 0·2-1·0 kg a.i. per hectare; and nematodes by both foliar and soil application.

Caution

1 This is a very poisonous substance (Part II p. 93) — full protective clothing should be worn.
2 Dangerous to fish, game and wild life.

Formulation Water soluble liquid 24 g a.i. per litre, granules 50 and 100 g a.i. per kilogram.

Primicarb
(2-dimethylamino-5,6-dimethylpyrimid-4-yl dimethylcarbamate)

Trade names 'Pirimor', 'Aphox'

Properties A selective insecticide used against Diptera. It is effective against organophosphorus-resistant strains. It is fast-acting with fumigant and translaminar properties — it is taken up by the roots and translocated in the xylem vessels. Non-acaricidal.

It is a colourless solid, virtually insoluble in water, but soluble in most organic solvents. Forms well-defined crystalline salts with acids; these salts are water-soluble.

The acute oral LD50 for rats is 147 mg per kilogram.

Use Effective against organophosphorus-resistant strains of aphids and dipterous maggots.

Formulation 50 per cent w.p. and 5 per cent granules.

Promecarb
(5-isopropyl-*m*-tolyl methylcarbamate)

Trade name 'Carbamult'

Properties A non-systemic contact insecticide, useful against coleopterous pests, Lepidoptera and Diptera.

It is a colourless crystalline solid, slightly soluble in water, soluble in propanone and dichlorethene; hydrolysed by alkali.

The acute oral LD50 for rats is 70 - 100 mg per kilogram.

Use Effective against Coleoptera, Lepidoptera, and fruit leaf-miners.

Caution This is a poisonous substance — protective clothing must be worn.

Formulation 25 per cent e.c.; 37·5 and 50 per cent w.p.; 5 per cent dust.

Propoxur
(2-isopropoxyphenyl metahylcarbamate)

Trade names 'Baygon', 'Blattanex', 'Unden'

Properties A non-systemic insecticide, with rapid knock-down, used against hemipteran flies, millipedes, ants and other household and public health pests. Non-phytotoxic.

It is a white crystalline powder, only slightly soluble in water, but soluble in most organic solvents; unstable in highly alkaline media.

The acute oral LD50 for rats is about 100 mg per kilogram.

Use Effective against jassids, aphids, other bugs, flies, millipedes, termites, ants and other household pests, mosquitoes, ticks and mites.

Caution

1 This is a poisonous substance — protective clothing should be worn.

2 Dangerous to bees.

Formulation To e.c.s.; w.p.s; dusts; granules; baits; and pressurised sprays of different a.i. concentrations; some sprays with added dichlorvos.

Miscellaneous compounds

There are organic fumigants and inorganic salts which possess toxic qualities to insect pests. The gases and fumigants act through the insect respiratory system, but the inorganic salts act usually as stomach poisons. Among these compounds are some of the earliest pesticides used in agriculture.

Aluminium phosphide

Trade name 'Phostoxin'

Properties Aluminium phosphide is a yellow crystalline solid, stable when dry, but reacts with moist air to liberate phosphine. Phosphine is highly insecticidal and a potent mammalian poison; it is spontaneously inflammable in air. The residue is harmless.

Use Fumigation of stored products and containers only.

Caution This is a very potent poison — fumigation takes from 3 - 10 days and should only be undertaken by trained personnel.

Formulation 'Phostoxin' evolves a non-inflammable mixture of phosphine, ammonia and carbon dioxide; manufactured as tablets (3 g) or pellets (0·6 g).

Bromomethane

Properties A general poison with high insecticidal and some acaricidal properties. It is used for space fumigation and for the fumigation of plants and plant products in stores. It can be used for the control of nematodes, fungi and weeds as a soil fumigant.

It is a colourless gas, scarcely soluble in water, soluble in most organic solvents; stable, non-corrosive and non-inflammable.

Use Effective against pests of stored products and of soil.

Caution A very poisonous gas, highly toxic to man — use should be restricted to trained personnel.

Formulation Packed as a liquid in glass ampoules (up to 50 ml), in metal cans and cylinders for direct use. Chloropicrin is sometimes added, up to 2 per cent as a warning gas.

Paris Green
(copper ethoxyarsenate(III))

Properties A stomach poison used as baits, very phytotoxic. Introduced about 1867 for the control of Colorado beetle.

It is a green powder of low solubility in water; in the presence of water and carbon dioxide it readily decomposes to give water-soluble and phytotoxic arsenical compounds.

Acute oral LD50 for rats is 22 mg per kilogram; a violent poison when ingested.

Use Effective only as baits because of high phytotoxicity.

Caution A very poisonous substance which should be handled with great care; a persistent and accumulative poison.

Formulation Most specifications require a content of at least 35 per cent arsenic(III) oxide, 20 per cent copper(II) oxide, 10 per cent ethanoic acid; and not more than 1·5 per cent arsenic(III) oxide in a water-soluble form.

Dibromoethane

Trade name 'Bromofume'

Properties An insecticidal fumigant used against stored products pests and for treatment of fruit and vegetables, and for soil treatment against certain insects and nematodes; very phytotoxic.

It is a colourless liquid, insoluble in water, but soluble in ethanol, ethoxyethane and most organic solvents; stable and non-inflammable.

The acute oral LD50 for male rats is 146 mg per kilogram; dermal application will cause severe burning.

Use Used against pests of stored products; for the treatment of fruit and vegetables and for soil treatment against certain insects and nematodes. If used for soil treatment planting must be delayed until 8 days after treatment because of its phytotoxicity.

Caution This is a dangerous substance and should be handled with care; protective clothing should be worn.

Formulation 'Dowfume W-85' per cent, for soil use in solution in an inert solvent.

Lead(II) arsenate(IV) ($PbHAsO_4$)

Properties A non-systemic stomach insecticide, with little contact action. It is non-phytotoxic, though the addition of components causing the production of water-soluble arsenical compounds may lead to leaf damage.

Formulated as either paste or powder, it is poisonous to mammals when ingested.

The acute oral LD50 for rats is 10–50 mg per kilogram.

Use Used against various caterpillars, sawfly larvae, and tipulid larvae. Does not give complete control of caterpillars, but is selective and does not harm predators of red spider mites.

Caution

1 Dangerous to bees.
2 Harmful to fish and livestock.
3 Pre-harvest interval for edible crops is 6 weeks.
4 Pre-access interval for livestock to treated areas is 6 weeks; 3 weeks in wet weather.

Formulation As wettable powders or pastes.

Mercury(I) chloride

Trade name Calomel

Properties A general poison, but being phytotoxic its use is generally limited to soil application, mainly for the control of root maggots. It is also used as a fungicide and for the control of club-root of *Brassica* spp. Its biological activity arises from its reduction to metallic mercury.

It is a white powder, insoluble in water, but soluble in ethanol and most organic solvents. In the presence of water it is slowly dissociated to mercury and mercury (II) chloride; the rate of this reaction is increased by alkalis.

The acute oral LD50 for rats is 210 mg per kilogram.

Use Moderately effective against soil-inhabiting root maggots of vegetables. Use generally restricted also by the high cost of the chemical.

Formulation The pure compound is used as a seed dressing; a 4 per cent dust on non-alkaline carrier.

Sulphur

Properties A non-systemic direct and protective fungicide and acaricide. Generally non-phytotoxic, except to certain varieties known as 'sulphur-shy'.

It is a yellow solid, existing in allotropic forms; practically insoluble in water, slightly soluble in ethanol and ethoxyethane. Slowly hydrolysed by water; compatible with most other pesticides, except petroleum oils.

Use Used mainly against phytophagous mites, especially against 'big-bud' mites on blackcurrants. Also effective against powdery mildews, and apple scab.

1 Some fruit varieties are 'sulphur shy' — follow the manufacturer's instructions on label for susceptible varieties.
2 To avoid possible taint, do not use on fruit for processing.

Formulation Dusts; w.p.; or as finely ground colloidal suspensions, or more usually as 'Lime-sulphur'.

Natural organic compounds.

Many plants contain toxic compounds. Some are selectively poisonous to insects and have proved valuable as insecticides. These compounds do not appear to induce resistance in the insect pests. However, pyrethrins are so useful that extensive research is now yielding synthetically produced pyrethrins, some of which are more effective than their naturally occurring analogues.

Bioallethrin
((±)-3-allyl-2-methyl-4-oxocyclopent-2-enyl(+)- *trans*-chrysanthemate)

Trade names 'Bioallethrine', 'D-Trans', 'Esbiol'

Properties A powerful contact insecticide, by nature a synthetic pyrethroid, causing a rapid knock-down. Metabolic detoxication is delayed by addition of synergists such as piperonyl butoxide. It is more persistent than natural pyrethrins. It occurs as 2 isomers, but both are very similar; insoluble in water, but miscible with most organic solvents.

The acute oral LD50 for male rats is 500-780 mg per kilogram.

Use Effective against most household pests, especially Diptera.

Formulation Mostly in combination with synergists and other insecticides in kerosene as fly sprays, aerosols (0·1-0·6 per cent a.i.), or in impregnated dusts.

Nicotine
((*S*-3-(1-methylpyrrolidin-2-yl)pyridine)

Properties A non-persistent, non-systemic, contact insecticide with some ovicidal properties. Can be used as a fumigant in closed spaces, water insoluble salts ('fixed' nicotines) have been used as stomach insecticides. Nicotine is prepared from tobacco (*Nicotiana tabacum*) by steam distillation of solvent extract.

A colourless liquid, darkening on exposure to air, miscible with water below 60 °C (forming a hydrate); miscible with ethanol, ethoxyethane, and readily soluble in most organic solvents.

The acute oral LD50 for rats is 50-60 mg per kilogram.

Use Effective against aphids, capsids, leaf-miners, thrips on a wide range of horticultural crops; sawflies and woolly aphid on apple. Also used for fumigation of glasshouses.

Caution

1 Harmful to bees.

2 Dangerous to fish, livestock, game, wild birds and mammals.

3 Pre-harvest interval for outdoor edible crops is 2 days.

4 Pre-access interval for livestock to treated areas is 12 hours.

Formulation Marketed as the 95 per cent alkaloid, or as nicotine sulphate(VI) (40 per cent alkaloids); also as 3-5 per cent dusts. For fumigation nicotine shreds are burnt, or the liquid nicotine is applied to a heated metal surface.

Pyrethrins

Properties Pyrethrin is a general term including the cinerins as well as the pyrethrins extracted from flowers of *Pyrethrum cineraefolium*. They are powerful contact insecticides; non-systemic; causing a rapid paralysis or knock-down. They are unstable to sunlight and are rapidly hydrolysed by alkalis with a loss of insecticidal properties. Their metabolic detoxication may be delayed by the addition of synergists such as piperonyl butoxide, sesamin, etc.

The acute oral LD50 for rats is about 250 mg per kilogram.

Use Effective against flies, and household pests, usually with added synergist.

Formulation As dusts with added non-alkaline carrier; as aerosols the extract is dissolved in a volatile solvent such as chloromethane or dichlorodifluoromethane, usually with added synergist.

Rotenone

Trade name 'Derris'

Properties This is the name given to the main insecticidal compound of certain *Derris* spp., and *Lonchocarpus* spp., known for many years to be effective as a fish poison and an insecticide. It is a selective non-systemic insecticide with some acaricidal properties; non-phytotoxic. It has low persistence in spray or dust residues. Insoluble in water but soluble in polar organic solvents; readily oxidised in presence of light and alkali to less insecticidal products.

The acute oral LD50 for rats is 132 - 1 500 mg per kilogram; it is very toxic to pigs.

Use Effective against aphids, caterpillars, thrips, some beetles and red spider mites.

Caution

1 Dangerous to fish.
2 Toxic to pigs.
3 Pre-harvest interval for edible crops is 1 day.

Formulation Usually as dusts of the ground root with a non-alkaline carrier; dusts may be stabilised by addition of a small quantity of a strong acid such as phosphoric(v) acid.

Organic or hydrocarbon oils

Oils distilled from crude mineral oils (petroleum or mineral oils) or from coal tar (tar oils). These oils are chemically complex. They are composed of a number of individual compounds. They are all strongly phytotoxic and use is generally restricted to dormant season washes, when they are destructive to insect eggs.

Petroleum oils

Trade name 'Volck'

Properties Also known as minerals oils; refined grades are known as white oils. The use of kerosene as an insecticide dates from about the time of its introduction as an illuminant. Oils distilling at higher temperatures came into use about 1922. They consist largely of aliphatic hydrocarbons both saturated and unsaturated. They are produced by distillation and refinement of crude mineral oils; the ones used as pesticides generally distil above 310 °C and may be classed as 'light', 'medium', or 'heavy' on the basis of the percentage distilling at 335 °C: 64 - 79, 40 - 49 and 10 - 25 per cent respectively; density rarely exceeds 0·92 at 15·6 °C. Viscosity and density vary according to the source of the crude oil.

Use Effective against certain insects such as mealybugs, scales and thrips, and against red spider mites; they are ovicidal. Their use is limited by their phytotoxicity. A semi-refined oil can be used as a dormant ovicide; for foliage use a refined oil of narrow viscosity range is required. Relatively harmless to mammals.

Formulation Available either alone or in mixture with DNOC or DNOC and DDT.

Tar oils

Properties Produced by distillation of tars resulting from the high-temperature carbonisation of coal and coke oven tars. Although used for wood preservation since 1890, the introduction of the formulated products known as tar oil washes for crop protection dates from about 1920. These oils are brown to black liquids, of density 1·05 - 1·11; insoluble in water but soluble in organic solvents. They consist mainly of aromatic hydrocarbons but contain phenols and tar acids. They are highly phytotoxic.

Use Effective for the control of the eggs of many insect species, particularly of aphids, Lepidoptera, psyllids and other bugs. Because of phytotoxicity use is restricted to the dormant season.

Caution

1 Dangerous to fish.
2 Irritating to skin, eyes, nose and mouth.
3 Phytotoxic.

Formulation As miscible winter washes, and stock emulsion winter washes. The most insecticidally active tar oil is creosote.

Biological compounds

A number of viruses and bacteria, and some fungi,

are recorded as causing diseases in insect populations. On several occasions when part of a large pest population has been found to be naturally infested by a pathogen it has been possible to make up a spray solution from a suspension of macerated, diseased insect bodies and to spray this solution over uninfected parts of the pest population, often with spectacular success. The insect groups that regularly suffer from viral and bacterial epizootics are Lepidoptera and Coleoptera.

The only biological compound at present commercially available for use against insect pests is a spore suspension of *Bacillus thuringiensis* for use against lepidopterous larvae only. Various strains of *Bacillus popilliae* cause Japanese beetle milky disease.

Bacillus thuringiensis Berliner

Trade names BTB-183 'Biotrol', 'Thuricide'.

Properties This biological compound is a suspension of spores in an inert powder, specific in action to lepidopterous larvae, on which it acts as a stomach poison only. There are no phytotoxic effects, and it is innocuous to mammals, other animals, and insect predators and parasites.

It is a stable product which has been kept at room temperature for over 6 years without any detectable loss in potency. However, since it is a viable biological compound it has to be protected from extremes of heat and light, and corrosive fumes. This bacterium, in addition to forming resistant spores, also produces a crystal of thermolabile endotoxin called a parasporal body. It is the parasporal body which is the insecticidal agent in any preparation of *B. thuringiensis*. The crystalline toxin is insoluble in water and is thus deposited intact upon the leaf surface. After ingestion by the caterpillar it acts as a stomach poison. It has no contact action whatsoever. Some caterpillars are killed within a few hours of feeding but other less susceptible species are not killed directly, they suffer a gut paralysis which stops them feeding and they die within a few days. A few important species of lepidopterous pests do not appear to be susceptible at all to this pathogen.

Use This is effective only against the larvae of Lepidoptera, but is not equally effective against all caterpillars.

Caution There appears to be no real restriction against the use of this insecticide at all. However, it is incompatible with those insecticides which cause large changes in pH.

Formulation It is formulated as a wettable powder containing 25 billion viable spores per gram of product (2·5 per cent), and 97·5 per cent of inert ingredients.

Diflubenzuron

Trade name 'DIMILIN'

Properties An insect growth regulator (IGR) produced in 1974, operating by ingestion, and interference with the deposition of cuticle at ecdysis. A species-dependent ovicidal contact action has been demonstrated in *Spodoptera littoralis*, as well as prevention of egg eclosion after uptake by females. It has no systemic action and so is not effective against sap-sucking insects. In soil it is rapidly degraded.

The technical product is pale brown crystals, insoluble in water and apolar solvents, but is soluble in polar solvents.

The acute oral LD50 for rats was more than 4 640 mg per kilogram.

Use Effective against a broad range of leaf-eating insects and some mites; at rates of 1·5-30 g a.i. per 100 litres. Also used against mosquito and fly larvae.

Formulation Liquid 10 per cent (8 g per litre 250 g a.i. per kilogram.

Methoprene

Trade name 'Altosid'

Properties An insecticide of the insect growth regulator (IGR) group. It is of short persistence in the field since it is rapidly biodegraded by micro-organisms and sunlight. No recorded damage to non-target organisms, especially predators.

It is a pale amber liquid, insoluble in water, but soluble in most organic solvents. Completely non-toxic to laboratory tested mammals.

Use Most effective against Diptera (larvae),

and developed initially for destruction of mosquito larvae. It is now used for systemic treatment of cattle against hornfly, and is being tested against various pests of stored products. Treated larvae develop apparently normally into pupae; but the pupae die without giving rise to adults.

Formulation Liquid 10 per cent (8 g per litre a.i.).

18 Materials used to control plant diseases

Sterilants

Heat can be used to sterilise soils on a small scale, e.g. for greenhouse beds, potting, etc. (p. 66). The general principles involved in applying chemicals to soil are discussed on p. 75. The chemicals used for soil sterilisation are general biocides and will kill nematodes, insects and plants as well as soil pathogens. Most are fairly poisonous and should be used with adequate precautions.

Bromomethane (methylbromide)

Properties A colourless, almost odourless, toxic gas.

Use As a general fumigant for dead plant products, closed buildings and to sterilise soil. The gas may be injected into the soil (which should be covered) using 1 kg (compressed) per 10 square metres. Alternatively, a solution of the gas, usually 20 per cent, is available which may be injected at 5 ml per litre of soil. The gas is an efficient fumigant at fairly low temperatures (10 °C) and disperses easily.

Caution

1 Highly toxic to animals, although less irritant than chloropicrin, and operators must wear respirators.
2 Less toxic to plants than chloropicrin.
3 Wait 7 days before planting.

Formulation Available as a liquid under pressure in ampoules, cans or cylinders for direct use. Chloropicrin, because of its characteristic odour, may be added in small quantities as a warning gas.

Chloropicrin (trichloronitromethane)

Properties A highly volatile irritant toxic liquid.

Use To sterilise soil by applying with an injection gun at about 25 cm spacing at a depth of 10 - 12 cm, using 3 ml doses. For field application, tractor-drawn injection equipment can be used. The rates used vary from 50 - 300 litres per hectare depending on soil conditions, equipment used, depth of injection, etc.

Small quantities of soil can be treated in closed containers, using not more than 1 ml per litre of soil. Soil should not be below 18 °C when treated. Sclerotia of some soil fungi are resistant.

Caution

1 This substance is very toxic and strongly irritant to skin, eyes and respiratory passages (mustard gas) — full protective clothing, including goggles and respirators, must be worn.
2 Pre-planting time is 2 weeks if soil is warm and well-ventilated; if conditions for dispersal are poor then a longer time should pass.

Formulation Available as the pure liquid in closed containers.

Dazomet (tetrahydro-3,5-dimethyl-1,3,5-thidiazine-2-thione)

Trade names 'Basamid', 'Mylone'

Properties A white crystalline solid which decomposes in soil to produce methyl isothiocyanate.

Use It is incorporated into soil at the rate of 300 - 400 kg per hectare. The powder should be thoroughly mixed with the soil using a rotary cultivator or similar machine and the treated area adequately irrigated.

Caution

1 Dazomet and its decomposition products are irritants and moderately toxic.
2 Wait 4 - 8 weeks before planting.

Formulation Available as a 85 per cent a.i. powder ('Basamid', 'Mylone') or 95 per cent granules.

Dibromoethene

Trade names 'Dowfume W-85', 'Bromofume', 'Doxlone'

Properties A colourless liquid, slightly soluble in water.

Use As a general soil fumigant.

Caution Moderately toxic and phytotoxic. Wait 1 - 2 weeks before planting.

Formulation As a solution in an inert solvent ('Dowfume W-85', 'Bromofume'); with dichloropropene ('Doxlone').

Metham-sodium (sodium *N*-methyl dithiocarbamate)

Trade name 'Vapam'

Properties A white crystalline solid with a pungent odour, readily decomposing to release a toxic gas, methyl isothiocyanate, which is the active principle.

Use As a drench to sterilise soil. A dilute (1:15) aqueous solution of the liquid is usually applied at the rate of 2·5 litres per square metre. The gas released on decomposition efficiently fumigates the soil at temperatures above 10 °C.

Caution

1 The fumigant is irritant, but is less toxic than other soil fumigants.
2 Wait 4 - 10 weeks before planting.

Formulation 'Vapam 4-s' is an aqueous solution of 32·7 per cent of the anhydrous sodium salt.

Methanal (formaldehyde)

Trade name 'Formalin'

Properties A colourless irritant gas.

Use A dilute (1·5) solution of 'Formalin' (40 per cent aqueous) is used as a general-purpose sterilant for pots, boxes, tools, etc., and is also applied as a drench at the rate of 2·5 litres per square metre to sterilise seedbeds, etc.

Caution

1 When used in warm, confined spaces, irritation of eyes and nose may be experienced as the gas evaporates.
2 Poisonous to man and animals.
3 Treated soil should be left for 1 - 2 weeks before planting.

Formulation Available as 40 per cent aqueous solution ('Formalin') and occasionally as the polymerised solid.

Methyl isothiocyanate

Trade names 'Trapex', 'Di-Trapex', 'Vorlex'

Properties Colourless volatile crystals with pungent odour, slightly soluble in water. Corrosive to metals.

Use As a general soil fumigant against fungi, insects, nematodes and weed seeds, applied at about 20 g per square metre of soil.

Caution

1 Moderately toxic, irritating to skin, eyes and respiratory passages.
2 Phytotoxic.
3 Soil must be left 3 weeks (warm weather) to 8 weeks (cool weather) to allow dispersal of the fumigant before planting.

Formulation As emulsifiable concentrate, and with dichloropropane-dichloropropene.

Fungicides and bactericides

Some of the chemicals which are active against fungal diseases may also be used against bacterial diseases and it is convenient to deal with these substances together. Fungicides also may be fungistatic, substances which stop growth or antisporulant, substances which prevent sporulation. Bactericides are similarly divided into bacteristatic and lethal substances. In contrast to sterilants, these substances are usually applied to the growing plant, so that they must not be phytotoxic at the rates used for disease control. Protective chemicals do not enter the plant tissue, they remain on the outside of the plant where they kill the growing fungal hyphae but do not damage the plant. Therapeutants or systemic fungicides must enter plant tissues to be effective. These substances are able to kill the pathogen without damaging the host cells. The physiological processes of fungus or bacterium and host cells in intimate parasitic association are very similar. It is difficult to produce a toxin for systemic use, which can take advantage of the few differences which exist, to kill the pathogen selectively. The toxin should be efficiently absorbed and translocated through the plant tissues. Few systemic fungicides do move efficiently through plants as they are transported only in the xylem and not in the phloem.

A wide variety of chemical compounds are used as fungicides. Each year some compounds are superseded by more efficient, cheaper or less hazardous compounds. No attempt is made here to give a comprehensive list of all the fungicides as this can be found in other publications. The major groups of substances used, and the more widely available compounds are reviewed.

Non-systemic fungicides

Copper-based fungicides were among the first chemicals to be used in crop protection and Bordeaux mixture (colloidal precipitate formed by mixing lime with copper(II) sulphate(VI) solution) is still one of the most efficient of these. Other mixtures using copper(II) sulphate(VI) include Burgundy (with sodium carbonate) and Cheshunt (with ammonium carbonate) mixtures. Proprietary formulations of copper(II) oxide, hydroxide and oxychloride have mostly superseded Bordeaux mixture. Other copper salts are used as wood preservatives. Copper fungicides are usually available as wettable powders and are applied at 0·1–0·5 per cent a.i. in water. Their persistence, broad spectrum of activity including bactericidal activity, which most other fungicides lack, and relative cheapness are valuable assets.

Mercury compounds are still valuable fungicides although they are being replaced by safer organic compounds. Mercury(I) chloride can be used as a soil fungicide, but most mercurial fungicides used today are formulated from organomercurial compounds such as methoxyethyl mercury salts and applied as seed dressings — they have a broad spectrum of activity with eradicant properties.

Sulphur is still used to some extent for powdery mildew control and in lime-sulphur against fruit tree diseases such as apple scab and peach leaf curl.

Organo-tin compounds, e.g. triphenyltin salts are used in some fungicides.

Dithiocarbamate compounds which were introduced in the 1930s were the first of the modern organic fungicides. These are derivatives of dithiocarbamic acid — usually metallic salts. Thiram, one of the first dithiocarbamates, is still widely used as a seed and soil fungicide. Zineb, maneb, mancozeb, metiram and propineb are used in various fungicides against a very wide range of diseases usually at 0·1–0·2 per cent a.i. spray or 1–3 kg a.i. per hectare.

A wide range of other broad-spectrum protectant organic fungicides exists including captan, a well-established fungicide often used in seed dressings. The related compounds captafol, chlorothalonil and folpet are widely used protectants which have some fumigant and curative action. Others are listed in the accompanying table.

Organic protectants with particular activity against a narrow range of fungal pathogens have been developed for some diseases. These include binapacryl, dinocap, quinomethionate against powdery mildews, dicarboximide compounds such as dichloran, iprodione and procymidone against *Botrytis* spp. and edifenphos against rice blast.

Some fungicides have been developed specially for seed and soil application, e.g. chloranil, quintozene and guazatine. Fenaminosulf and etridiazole are particularly effective against soil Oomycetes.

Systemic fungicides

Since the 1960s there has been a substantial

development of these compounds. Many have been based on or are closely related to methyl benzimidazole carbamate (MBC) and related compounds. Benomyl, carbendazim, thiabendazole and thiophanate are the main examples. These are effective against most Ascomycetes and Fungi Imperfecti but have little effect on Oomycetes and Basidiomycetes.

Pyrimidine compounds, e.g. ethirimol, dimethirimol and bupirimate are used against powdery mildews. Carboxylic acid anilides such as benodanil and carboxin are particularly effective against Basidiomycetes. Chloroneb, a widely used systemic seed dressing, is a chlorinated hydrocarbon. Dodemorph and tridemorph are morpholine compounds with a fairly wide spectrum of activity. Triazole compounds such as triadimefon, diclobutrazol which are very effective against a wide range of diseases on many crops are among the more recent developments. Acylalanine compounds such as metalaxyl and furalaxyl show a high level of activity against Oomycetes as does aluminium *tris*(ethyl phosphonate).

Antibiotics usually act systemically and some are used as fungicides or bactericides. Blasticidin and kasugamycin are used against rice blast in Japan; formulations of cycloheximide and streptomycin can be used on high-value cash crops against bacterial diseases. Tetracycline antibiotics can cause remission of symptoms in diseases caused by mycoplasma-like organisms.

Resistance, or more accurately, tolerance has developed to many systemic fungicides among populations of fungal pathogens especially when they have been exposed to a high level of selection pressure (e.g. prolonged exposure) and where the fungicide used has a specific, single site, action. Pathogens have developed resistance most quickly to the MBC compounds, and to some of the pyrimidine compounds.

The fungicides have been classified in a number of ways, e.g. based on their chemical structure. A classification based on usage is probably of more practical benefit, and the three main types of use are listed on p. 74 et seq. In the following survey chemicals are grouped according to whether they are primarily inorganic or organic in structure, or are antibiotics; whether they are protectant, or systemic (eradicant) in action; and whether they are applied primarily to seed, soil or standing crops.

Inorganic fungicides: Copper compounds

Copper-based fungicides were among the first crop protection chemicals to be used and are still used very extensively. Bordeaux mixture was the foundation on which chemical disease control was built. It is still used today, usually as ready-mixed 'instant' spray. Other similar compounds (copper(II) sulphate(VI) mixed with an alkaline base) are used, but most readily available and easily used copper fungicides now contain simple basic copper salts. Several proprietary formulations use mixtures of these basic salts.

Copper fungicides have low mammalian toxicity and apart from the general precautions used when handling pesticides no special precautions are needed. The rate of application of copper fungicides depends on the formulation and on the crop on which they are used, but generally a dilute suspension in water (0·1 – 0·5 per cent of the active ingredient) is applied at medium or high volumes to ensure adequate coverage of the foliage. Actual amounts used per hectare vary with the crop.

Bordeaux mixture

Properties An amorphous flocculent blue precipitate of copper(II) hydroxide and calcium(II) sulphate(VI). It is corrosive to metals and incompatible with most other pesticides.

Use A broad-spectrum protectant fungicide for foliage with high tenacity if used immediately after mixing. Application at medium to high volume of the recommended formulation ensures adequate coverage.

Caution Can be phytotoxic to some plants under certain conditions; low mammalian toxicity.

Formulation May be prepared on site by mixing 15 parts (by weight) of calcium(II) hydroxide (hydrated lime) with a solution of 10 parts copper(II) sulphate(VI) ('Blue stone') in 100 parts of water. Some pre-mixed proprietary brands which only require the addition of water are also available. 'Macuprax' is pre-mixed Bordeaux and 'Cufraneb'.

Burgundy mixture

This is similar to Bordeaux mixture, except that sodium carbonate replaces the calcium(II) hydroxide to give a blue precipitate of basic copper(II) carbonate. Pre-mixture powders, e.g. 'Burcop' are available and often used as 'Bordeaux substitutes'. Copper(II) carbonate (malachite) has been used as a cereal seed dressing.

Cheshunt compound.

Properties This is a deep violet blue powder or suspension of complex cuprammonium carbonate and sulphate(VI). Slightly corrosive to some metals.

Use Broad-spectrum protectant fungicide used mostly as a soil drench or for application to pots, boxes, etc.

Caution Slightly phytotoxic with low mammalian toxicity and toxic to fish.

Formulation Can be prepared by mixing 2 parts of copper(II) sulphate(VI) with 11 parts of ammonium carbonate. Also available as proprietary soluble powders.

Copper(II) hydroxide

Trade names 'Kocide', 'Kocide 202'.

Properties Insoluble pale blue powder.

Use Similar to the other protectants; usually used at a rate of 0·1–0·5 per cent a.i. at high-volume spray. Kocide 202 is for low-volume application.

Caution Generally non-phytotoxic with low mammalian toxicity.

Formulation Wettable powder of 80–90 per cent a.i.

Copper naphthenates

Trade name 'Cuprinol'

Properties Green viscous liquid, immiscible with water, unpleasant odour, inflammable.

Caution Phytotoxic; low mammalian toxicity.

Use As a wood or fabric preservative applied neat as dip or paint.

Formulation Available as neat liquid having various copper concentrations up to 8 per cent. It may be formulated with other preservative substances such as oils, creosote, etc.

Copper(I) oxide

Trade names 'Perenox', 'Cuprocide'

Properties Insoluble red to yellow powder.

Use Broad-spectrum of activity, used at a rate of 0·2–0·5 per cent a.i. for adequate foliage cover (spray to run off). Can be applied to seed.

Caution Generally non-phytotoxic with low mammalian toxicity.

Formulation Usually as a wettable powder of 50 per cent a.i., but dusts and emusifiable concentrates also available.

Copper oxychloride

Trade names 'Vitigran', 'Blitox', 'Cupravit'

Properties Insoluble, pale blue-green powder. Corrosive to iron and steel in high concentrations.

Caution It is generally non-phytotoxic, with low mammalian toxicity.

Formulation Usually as a wettable powder (50 per cent a.i.) also as 10 or 25 per cent dusts.

Copper(II) sulphate(VI)

Trade names 'Blue Stone', 'Blue vitriol'

Properties Blue crystals or powder, soluble in water.

Use As an algicide and wood preservative or for making Bordeaux mixture, etc.

Caution Low phyto- and mammalian toxicity but toxic to fish.

Formulation As undiluted compound for use in solution and in other copper fungicides.

Oxine-copper (copper hydroxyquinolate)

Properties Insoluble, chemically inert greenish yellow powder.

Use Mildew proofing of fabric, application to boxes, pots, staging, etc. Has also been used occasionally as a foliar protectant and as a seed dressing.

Caution Not allowed to be used on food crops in USA.

Formulation Usually formulated as a wettable powder but liquid preparations are also available.

Inorganic fungicides: Sulphur compounds

Lime-sulphur

Properties Deep orange liquid, with unpleasant odour, a solution of calcium polysulphides.

Use Mainly as a substitute for sulphur for foliar application against powdery mildews, peach leaf curl and some other diseases, at about 1 per cent dilution in water.

Caution Can cause scorching; do not use on sulphur-shy plants, disagreeable to use, caustic, corrosive, moderately toxic and causes skin irritation. Non-compatible with many other pesticides.

Formulation Usually supplied as undiluted liquid for dilution in water.

Sulphur

Properties Yellow powder or crystals.

Use Sulphur is an important fungicide for the control of powdery mildews, and is usually applied as a finely ground dust of elemental sulphur or as the 0·1–0·2 per cent suspension of a water-dispersible powder.

Caution Do not use during high temperature or with oils. Sulphur is phytotoxic to cucurbits and certain sulphur-shy varieties of other crops and cannot be used on them. Non-toxic to mammals, but may be irritant.

Inorganic fungicides: Mercurial compounds

Mercurial fungicides have a very powerful eradicant action on many diseases. They present an environmental hazard because they are phytotoxic and toxic to animals. The organo-metallic compounds are the least phytotoxic mercurial compounds; these are used in crop protection where possible. Organo-mercurial compounds are chiefly used in seed dressings; they are rarely applied to standing crops. Some seed-borne pathogens have developed tolerance to mercury. For this reason and because of environmental hazards, organo-mercurial seed dressings are being replaced by safer compounds.

Mercury(I) and mercury(II) chloride

Other names Calomel and corrosive sublimate respectively

Properties White powders; the mercury(II) salt is fairly soluble in water. Mercury(I) chloride is less toxic and is used more frequently than mercury(II) chloride.

Use Soil and root pests and diseases, also applied to turf for moss and disease control.

Caution These compounds are general poisons and should be used with extreme care. Both are phytotoxic.

Formulation Used in high dilutions only, e.g. in lawn seed.

Methoxyethyl mercury salts

Properties White powders, fairly insoluble in water.

Use Seed dressings, especially on cereals; rates depend upon seed and formulation used. Has a powerful eradicant effect on deep-seated infections.

Caution Phytotoxic, toxic to man and wildlife.

Formulation Seed dressing powders or slurries containing about 2 per cent mercury, e.g. 'Ceresan', 'Aretan', often formulated with other pesticides.

Phenyl mercury acetate (PMA) and chloride (PMC)

Properties White solids; PMA is slightly soluble in water. Can be phytotoxic on some crops. Incompatible with many other fungicides.

Use Has been used as an eradicant spray (at

0·05 – 0·1 per cent a.i.) on high-value fruit crops. Also used to control turf diseases and as a seed dressing. Now largely superseded by safer chemicals.

Caution Toxic, and should be used with care.

Formulations As wettable powder and liquids.

Other organo-mercurial compounds.

Fungicides with similar action include
1 ethoxyethylmercury hydroxide ('Tillex');
2 ethylmercury *p*-toluene sulphonanilide ('Granosan M', 'Ceresan M');
3 methylmercury dicyandiamide ('Panogen').

Inorganic fungicides: Non-mercurial metallic compounds.

Triphenyltin compounds

Trade names 'Brestan' (fentin acetate), 'Du-Ter' (fentin hydroxide), 'Stannoram' (decafentin)

Properties White powders, insoluble in water.

Use Protectant fungicides with some eradicant action effective at low rates (0·03 – 0·05 per cent a.i. suspension) against many foliar pathogens. Antifeedant effect on insects. Tributyl tin oxide (TBTO) has been used as a wood preservative.

Caution
1 Generally low phytotoxicity.
2 Moderately toxic to fish and wildlife.
3 Do not use immediately before harvest.

Formulation Wettable powders or dusts.

Methyl arsenic sulphide

Trade names 'MAS', 'Rhizoctol', 'Urbasulf'

Properties Colourless flakes with unpleasant odour; insoluble in water.

Use Seed dressing on a variety of crops, especially against *Rhizoctonia* spp. Rate depends upon seed.

Caution Effectiveness influenced by environmental conditions. Can cause skin irritation, moderate mammalian toxicity.

Formulation It may be mixed with other fungicides to produce seed dressing dusts or slurries as in 'Rhizoctol' and 'Urbasulf'.

Cadmium, zinc and nickel salts.

These compounds have occasionally been used in fungicide formulations.

Organic fungicides: General protectants

Dithiocarbamates

The dithiocarbamate fungicides which are derivatives of dithiocarbamic acid usually as a metallic salt, were some of the first organic fungicides to be developed and are still widely used today. They are protectants with a broad spectrum of activity and are generally used at 0·2 – 0·1 per cent a.i. spray or 1 – 3 kg a.i. per hectare (depending upon the crop). They are applied at moderate to high volume to ensure adequate coverage.

Cufraneb

Properties An off-white powder composed of a complex dithiocarbamate containing copper, manganese, iron and zinc.

Use A broad-spectrum protectant fungicide used at 0·1 – 0·3 per cent a.i. in suspension.

Caution Low mammalian toxicity; can be irritating to skin, eyes, and nose. Do not use less than 1 week before harvesting.

Formulation As a wettable powder mixed with a Bordeaux mixture formulation as 'Macuprax'.

Ferbam
(ferric dimethyldithiocarbamate)

Trade name 'Fermate'

Properties Black powder virtually insoluble in water. Black deposit may be objectionable.

Use A protective fungicide for foliage application at 0·2 – 0·5 per cent a.i. suspension, now mostly superseded by other compounds.

Caution Non-phytotoxic and very low mammalian toxicity; can be irritating to skin, eyes and nose. Do not use less than 1 week before harvesting. Incompatible with copper, mercury and lime compounds.

Formulation Usually as wettable powder.

Mancozeb
(maneb/zinc complex)

Trade name 'Dithane M-45'
Properties A greyish-yellow powder, insoluble in water. Do not store in moist conditions.
Use Broad-spectrum foliar protectant used at 0·1 – 0·3 per cent a.i. suspension.
Caution Non-phytotoxic, very low mammalian toxicity; can be irritating to skin, eyes and nose. Do not use less than 1 week before harvesting.
Formulation Usually as 80 per cent a.i. wettable powder. 'Dikar' is mancozeb and dinitrophenyl mite repellant.

Maneb
(manganese 1,2-ethanediylbis (carbamodithioate))

Trade names 'Manzate', 'Dithane M-22'
Properties Yellowish crystalline powder slightly soluble in water. Decomposes when stored under moist conditions.
Use Widely used, broad-spectrum, foliar protectant at 0·1 – 0·3 per cent a.i. suspension. Sometimes applied to seed.
Caution
1 Toxic to fish.
2 Low mammalian toxicity; can be irritating to skin, eyes and nose; do not use close to harvest.
3 Usually non-phytotoxic.
Formulation Usually as wettable powders (80 per cent a.i.) or dusts.

Metiram
(ammonia complex of zinc ethylene thiuram disulphide))

Trade name 'Polyram-Combi'
Properties A yellowish powder; insoluble in water.
Use Broad-spectrum foliar protectant used at 0·2 – 0·3 per cent a.i. spray. Long residual effect.
Caution Can be used up to harvest. Non-phytotoxic with very low mammalian toxicity.
Formulation Wettable power (80 per cent a.i.) and dusts.

Propineb
([[(1-methyl-1,2-ethanediyl)bis[carbamodithioato]] (2-)] zinc homopolymer)

Trade name 'Antracol'
Properties Whitish powder, insoluble in water.
Use Broad-spectrum, foliar protectant fungicide with pronounced residual activity. Has some effect on mites and powdery mildews. Use at 0·2 – 0·3 per cent a.i. suspension.
Formulation As 65 per cent a.i. wettable powder and dusts.

Thiram
(tetramethylthioperoxydicarbonicdiamide)

Trade names 'Arasan', 'Tersan', 'Nomersan'
Properties Colourless crystals, virtually insoluble in water.
Use Widely used as a soil and seed fungicide and formerly as a foliar protectant. Has some animal repellent properties.
Caution Low mammalian toxicity, but may cause skin irritation. Non-phytotoxic, but do not use close to harvest. Can taint produce intended for food.
Formulation Usually available as a wettable powder or dust for seed or soil application.

Zineb
([[1,2-ethanediylbis[carbamodithioato]] (2-)zinc complex)

Trade names 'Dithane Z-78', 'Polyram-Z'
Properties White powder, insoluble in water.
Use A widely used broad-spectrum foliar protectant fungicide, also sometimes used for seed and soil application. Used at 0·2 – 0·5 per cent a.i. spray.
Caution Non-phytotoxic except to zinc-sensitive plants, very low mammalian toxicity; can be irritating to skin, eyes and nose. Do not use less than 1 week before harvesting.

Formulation As wettable powder (75 per cent usually) or dusts.

Ziram
((1-4)-bis(dimethylcarbamodithioato-S, *S'*)zinc)

Trade names 'Milban', 'Zerlate', 'Cuman'
Properties White powder, virtually insoluble in water.
Use A protective fungicide for foliar applications at 0·2–0·5 per cent a.i. suspension. Low persistence; now largely superseded by other compounds.
Caution Low mammalian toxicity; can be irritating to skin, eyes and nose. Non-phytotoxic except on zinc-sensitive plants. Do not use close to harvest.
Formulation Wettable powder and dusts.

Other broad-spectrum protectants

A large number of other organic compounds have also been used as protectants against a wide range of diseases.

Anilazine
(4,6-dichloro-N-(2-chlorophenyl)-1,3,5-triazin-2-amine)

Trade names 'Dyrene,' 'Triazine'
Properties A triazine; pale crystals; insoluble in water.
Use A protective fungicide used at about 0·2 per cent a.i. against pathogens such as *Botrytis* spp., *Septoria* spp. and *Colletotrichum* spp.
Caution Fairly low toxicity. Toxic to fish. Incompatible with alkaline compounds.
Formulation 50 per cent wettable powder.

Captafol
(3a,4,7,7a-tetrahydro-*N*-(1,1,2,2,-tetrachloroethanesulphenyl)phthalimide)

Trade name 'Difolatan'
Properties A chlorinated phthalimide; a whitish powder; insoluble in water and with a characteristic musty odour.
Use Protectant fungicide closely related to captan, used at 0·1–0·4 per cent a.i. suspension. Good persistence.
Caution Can be irritating to skin, eyes and nose. Toxic to fish. It can be phytotoxic to some crops during high humidities and high temperatures. Do not use less than 2 weeks before harvest.
Formulation 50 and 80 per cent a.i., wettable powders.

Captan
(3a,4,7,7a-tetrahydro-*N*-(trichloromethanesulphenyl)phthalmide)

Trade name 'Orthocide'
Properties A chlorinated phthalimide; a yellowish solid; insoluble in water and with a pungent odour.
Use A well-established and widely used protectant fungicide for foliage, seed and soil application, effective against a wide range of pathogens. Use at 0·2–0·5 per cent a.i. suspension, as a dust to soil before planting, as a seed dressing (rates depend on seed).
Caution Low mammalian toxicity, but may cause skin irritation, usually non-phytotoxic. Toxic to fish.
Formulation As wettable powders (50 and 80 per cent a.i.)

Chlorothalonil
(2,4,5,6-tetrachloro-1,3-benzenedicarbonitrile)

Trade names 'Daconil 2787', 'Termil', 'Bravo'
Properties White powder with slight odour; practically insoluble in water.
Use Protectant fungicide with broad-spectrum activity against foliar diseases and for sublimation in greenhouses.
Caution Toxic to fish. Very low mammalian toxicity; can be irritating to skin, eyes and nose. Low phytotoxicity; do not use close to harvest.
Formulation 75 per cent wettable powder; tablets for sublimation (smoke generation) in greenhouses.

Dichlofluanid
(*N*-dichlorofluoromethanesulphenyl-*N'N'*-dimethyl-*N*-phenylsulphamide)

Trade names 'Elvaron', 'Euparen'

Properties White powder; insoluble in water.

Use Protective fungicide widely used to control fruit tree diseases and has some effect on powdery mildews and mites.

Caution Low mammalian toxicity. Toxic to fish. Do not mix with alkaline substances. Low phytotoxicity; do not use less than 3 weeks before harvest.

Formulation 50 per cent wettable powder or dust.

Ditalimfos
(*O,O*-diethyl phthalimidophosphonothioate (I))

Trade name 'Plondrel'

Properties White crystals with a mild odour, slightly soluble in water.

Use Protectant and non-systemic curative fungicide for control of powdery mildews and apple scab. Applied at 0·3–0·5 per cent a.i. spray.

Caution Low mammalian toxicity but may cause skin irritation. Toxic to fish. Do not use less than 4 weeks from harvest.

Formulation 50 per cent wettable powder and 20 per cent emulsifiable concentrate.

Dithianon
(2,3-dicyano-1,4-dithia-anthraquinone)

Trade name 'Delan'

Properties Brown crystalline powder; insoluble in water. Incompatible with oils or alkalis.

Use Protective fungicide widely used to control many foliage diseases of fruit and ornamentals. Used at about 0·1 per cent a.i. spray.

Caution Low mammalian toxicity and not phytotoxic.

Formulation Available as a 75 per cent a.i. wettable powder and as a 25 per cent emulsifiable concentrate.

Dodine
(dodecylguanidinium acetate)

Trade names 'Cyprex', 'Melprex'

Properties White crystals soluble in water. It is a cationic surfactant.

Use A foliar protectant fungicide with some eradicant properties, widely used on fruit trees at a rate of 0·5–0·1 per cent a.i. (e.g. against apple scab).

Caution Low mammalian toxicity but irritant to skin; can be slightly phytotoxic at high concentrations to some crops.

Formulation Wettable powders, liquids and dust.

Drazoxolon
(4-(2-chlorophenylhydrazono)-3-methyl-5-isoxazolone)

Trade names 'Mil-Col', 'Ganocide'

Properties Yellow crystalline solid; almost insoluble in water.

Use Controls powdery mildews and some other foliage diseases on fruit trees. Also used as a seed dressing for control of seed-borne diseases and applied to collar region of trees to control root diseases (e.g. *Ganoderma* spp. on rubber).

Caution This is a Part III substance and full protective clothing should be worn. Dangerous to fish and livestock. Do not use less than 4 weeks before harvest.

Formulation Aqueous suspension ('Mil-Col') and a 10 per cent grease ('Ganocide').

Fenarimol
(2,4′-dichloro-α-(pyrimidin-5-yl)diphenylmethanol)

Trade names 'EL-222', 'Rimidin'

Properties White crystals; hardly soluble in water. Low toxicity.

Use Protectant and eradicant for use against diseases of fruit and ornamentals. Has limited systemic activity.

Caution Toxic to fish. Do not use close to harvest.

Formulation 4 and 12 per cent emulsifiable concentrate; 6 per cent wettable powders.

Folpet
(*N*-(trichloromethanesulphenyl)phthalimide)

Trade name 'Phaltan'
Properties White powder, insoluble in water.
Use Foliar protectant, used at 0·1 – 0·3 per cent a.i. in suspension.
Caution Generally considered non-toxic and can be used for post-harvest applications. Toxic to fish. Some phytotoxicity has been noticed in dry weather. Do not mix with oil-based or alkaline sprays.
Formulation Usually as a wettable powder.

Halacrinate
(7-bromo-5-chloroquinolin-8-yl acrylate)

Trade name 'Tilt'
Properties Brown crystals; very slightly soluble in water.
Use Non-systemic and curative fungicide for cereals. Applied at 1-2·5 per cent a.i.
Caution Toxic to fish. Irritating to skin and eyes.
Formulation As a wettable powder, 20 per cent a.i. with captafol (40 per cent).

Prochloraz
N-[2-(2,4,6-trichlorophenoxy)ethyl l]-N-propyl-2H-imidazole-1-carboximide)

Properties White crystalline solid; practically insoluble in water.
Use A broad-spectrum protectant fungicide with eradicant properties.
Formulation 25 and 40 per cent a.i. emulsifiable concentrates.

Pyridinitril
(2,6-dichloro-4-phenylpyridine-3,5-carbonitrile)

Trade name 'Ciluan'
Properties Colourless crystals; insoluble in water.
Uses Protectant fungicide for diseases of fruit and vegetables. Applied at 1 per cent a.i.
Caution Do not use less than 1 week before harvesting. Do not mix with alkaline spray.
Formulation As a 75 per cent w.p. and as 15 per cent pyridinitril with 25 per cent captan.

Tolylfluanid
(*N*-dichlorofluoromethanesulphenyl-*N'N'*-dimethyl-*N-p*-tolylsulphamide)

Trade name 'Euparen-M'
Properties Yellowish white powder; insoluble in water.
Use Similar to dichlofluanid, mainly used on deciduous fruit trees, against powdery mildews and as a mite suppressant.
Caution Low mammalian toxicity; toxic to fish. Low phytotoxicity. Do not use less than 3 weeks from harvest.
Formulation 50 per cent wettable powder.

Organic fungicides: Specific narrow-range protectants

These fungicides have specific action against certain diseases such as mildews and rice blast. These diseases are more difficult to control with the general protectants.

Binapacryl
(2-*sec*-butyl-4,6-dinitrophenyl-3-methylcrotonate (I))

Trade names 'Morocide', 'Endosan', 'Acricid'
Properties White powder with slight aromatic odour; insoluble in water.
Use Contact and protectant fungicide effective against powdery mildews also used against mites. Widely used on fruit crops at 0·05 – 0·1 per cent a.i.
Caution
1 Can be phytotoxic to young vegetable and ornamental crops.
2 Do not use less than 1 week before harvest.
3 Toxic to fish and livestock.
Formulation Wettable powders, dusts, aqueous and emulsifiable concentrate.

Chlorquinox
(5,6,7,8-tetrachloroquinoxaline)

Trade name 'Lucel'
Properties White powder; practically insoluble in water.

Use Active against powdery mildews, used on cereals; it has some systemic and eradicant activity. Applied at medium volume at 0·05–0·1 per cent a.i.

Caution Low mammalian toxicity.

Formulation As a wettable powder (25 per cent a.i.)

Dicloran
(2,6-dichloro-4-nitroaniline (I))

Trade names 'Botran', 'Allisan'

Properties Yellow solid; insoluble in water.

Use Specifically against fruit rots caused by *Botrytis* spp., *Rhizopus* spp., *Sclerotina* spp., etc. Applied as a high-volume dilute spray or dust to soil.

Caution

1 Can be phytotoxic on seedlings.
2 Very low mammalian toxicity.
3 Can be used up to and after harvest.

Formulation Wettable powder and dusts.

Dinobuton
(2-*sec*-butyl-4,6-dinitrophenylisopropyl carbonate (I))

Trade names 'Acrex'

Properties Pale yellow crystals; insoluble in water.

Use Contact fungicide for powdery mildews and an acaricide used at 5 per cent a.i.

Caution Toxic to fish. Do not use less than 1 week before harvest.

Formulation 30 per cent emulsifiable concentrate.

Dinocap
(isomeric reaction mixture of 2,6- dinitro-4-octylphenyl crotonates)

Trade names 'Karathane', 'Crotothane'

Properties A dinitrophenyl compound; dark brown liquid; insoluble in water.

Use A contact and persistent fungicide. Widely used to control powdery mildews and red spider mites. Applied at 0·05–0·2 per cent a.i.

Caution

1 Low mammalian toxicity; can be irritating to skin, eyes and nose.
2 Toxic to fish.
3 Non-phytotoxic except at high temperatures or with soil.
4 Do not use less than 1 week before harvest.

Formulation Available as emulsifiable concentrate, wettable powder and dusts.

Edifenphos
(*O*-ethyl-*SS*-diphenyl-phosphorodithioate)

Trade name 'Hinosan'

Properties Yellowish liquid with characteristic odour, insoluble in water. Incompatible with alkaline material.

Use Protectant with some eradicant action against rice blast and other rice diseases. Applied at 0·02–0·05 per cent a.i.

Caution Toxic to fish moderately so to mammals. Not compatible with alkaline material.

Formulation Emulsifiable concentrates and dusts.

Fluotrimazole
(1-(3-trifluoromethyltrityl)1,2,4-triazole)

Trade name 'Persulon'

Properties A colourless crystalline solid slightly soluble in water.

Use Protectant fungicide against powdery mildews.

Caution Can be irritating to skin and eyes, harmful to fish.

Formulation Wettable powder and emulsifiable concentrate.

Iprodione
(3-(3,5-dichlorphenyl)-1-isopropylcarbamoylhydantoin)

Trade name 'ROVRAL'

Properties White crystals; hardly soluble in water but soluble in organic solvents.

Use As a contact fungicide against *Botrytis* spp., *Monilia* spp. and *Sclerotina* spp. on fruit and as a seed dressing.

Caution Harmful to fish. Do not use less than 1 week before harvesting.

Formulation 'ROVRAL' is a 50 per cent w.p.

Nitrotal-isopropyl
(di-isopropyl-5-nitroisophthalate)

Trade names 'Kumulan' (with sulphur), 'Pallinal' (with Zineb)

Properties Yellow crystalline solid practically insoluble in water.

Use Protectant fungicide with specific activity against powdery mildews.

Caution Dangerous to fish and birds.

Formulation Wettable powders.

Petroleum oils

Properties Mixture of aliphatic hydrocarbons distilling above 310 °C. Flammable, phytotoxic.

Use Several types of mineral oils have been used as fungicidal sprays. Their general use is limited by phytotoxicity which depends upon the type of oil ('light', 'medium' or 'heavy'), the crop, application method and climatic conditions. Banana spray oil, used to control Sigatoka disease of bananas, is used throughout the Tropics, but apart from this, oils have achieved little popularity as fungicides although they are often used as additives either to enhance fungicidal activity, to act as spreaders and stickers, or as carriers of chemicals used in ULV formulations.

2-Phenylphenol

Trade names 'Dowicide 1', 'Dowicide A'

Properties White or pink crystals with mild odour; slightly soluble in water, the sodium salt ('Dowicide A') is more soluble.

Use Post-harvest fungicide used to control fruit mould, e.g. on citrus as 0·5 - 2 per cent dip. Also used to impregnate paper wraps, crates, etc., as a 5 per cent solution.

Caution Low mammalian toxicity but may cause skin irritation. Toxic to fish.

Formulation Wettable powder, wax flakes or liquid.

Procymidone
((*N*-3,5-dichlorophenyl)-1,2, dimethylcyclopropane-1,2-dicarboximide)

Trade name 'Sumisclex'

Properties A dicarboximide, white crystalline solid, slightly soluble in water.

Use Protectant fungicide against *Botrytis* spp. *Sclerotina* spp. and similar pathogens.

Caution Toxic to fish.

Formulation 50 per cent a.i. wettable powder.

Quinomethionate
(6-methyl-2-oxo-1,3,-dithiolo-[4,5,6] quinoxaline)

Trade name 'Morestan'

Properties A quinoxalin compound, yellow crystals practically insoluble in water.

Use Protectant fungicide used against powdery mildews — has acaricidal properties.

Caution Toxic to fish, may irritate skin.

Formulation 25 per cent a.i. wettable powder and 2 per cent dust.

Tecnazene
(1,2,4,5-tetrachloro-3-nitrobenzene)

Trade names 'Fusarex', 'Folosan'

Properties Volatile colourless crystals; insoluble in water.

Use Selective fungicide for *Fusarium* spp. and *Botrytis* spp. on potatoes and horticultural crops.

Formulation Emulsifiable concentrate (6 per cent a.i.), dusts, granules and smoke generators (often combined with BHC).

Thioquinox
(2-thio-1,3-dithiolo [4,5*b*] quinoxaline)

Trade name 'Eradex'

Properties Brownish powder; insoluble in water.

Use Contact and protectant fungicide against powdery mildews and mites.

Caution Low mammalian toxicity but can cause skin irritation.

Formulation As a wettable powder.

Vinclozolin
(3-(3,5-dichlorophenyl)-5-methyl-5-vinyl-1,3-oxazolidine-2,4-dione)

Trade name 'RONILAN'
Properties White crystals; slightly soluble in water.
Use Selective fungicide for control of *Sclerotina* spp., *Botrytis* spp. and *Monilia* spp. on fruit and vegetables at 1 kg a.i. per hectare.
Formulation 50 per cent a.i. wettable powder.

Protectant fungicides for seed and soil application

Included here are non-systemic organic compounds used chiefly to control seed or soil-borne diseases. Many of these have been promoted as safer substitutes for mercurial seed dressings. General fungicides such as thiram and captan may be used as seed or soil fungicides, but are also used on standing crops.

Bronopol
(2-bromo-2-nitro-1,3-propanediol)

Properties Pale solid; moderately (25 per cent) soluble in water.
Use A bacteriostat used to control seed-borne bacterial pathogens, e.g. blackarm disease of cotton.
Caution Moderate mammalian toxicity.
Formulation As seed dressing.

Chloranil
(2,3,5,6-tetrachloro-1,4-benzoquinone (I))

Trade name 'Spergon'
Properties Yellow crystals; virtually insoluble in water.
Use Mainly used for vegetable seed against seed- and soil-borne diseases.
Caution Do not use treated seed for food.
Formulation Seed dressing containing about 97 per cent a.i.

Dichlone
(2,3-dichloro-1,4-naphthoquinone (I))

Trade names 'Phygon', 'Uniroyal'
Properties Yellow subliming crystals; insoluble in water.
Use Mainly as a seed protectant but also as a foliage spray on fruit trees at 0·03 per cent a.i. and as an algicide.
Caution Toxic to fish. Do not use treated seed for food.
Formulation Wettable powder, paste and dusts.

Etridiazole
(5-ethoxy-3-trichloromethyl-1,2,4-thiadiazole)

Trade names 'Terrazole', 'Aaterra'
Properties White solid; insoluble in water.
Use Soil and seed application to control some common soil-borne root rots, damping-off fungi including *Pythium* spp.
Caution Low mammalian toxicity; can be irritating to skin.
Formulation Dust, granules, wettable powder.

Fenaminosulf
(sodium 4-dimethylaminobenzenediazosulphonate)

Trade names 'Dexon'
Properties Yellowish powder; slightly soluble in water.
Use On seed or in soil to protect particularly against Oomycete fungi, e.g. *Pythium* spp., *Phytophthora* spp.
Caution Harmful to birds and wildlife. Do not use treated seed for food.
Formulation Wettable powder and granules.

Fenfuram
(2-methyl-3-furanilide)

Properties White crystalline solid; slightly soluble in water.
Use Controls bunts and smuts of cereals, including those which are borne inside the seed.
Caution Do not use treated seed for food.
Formulation Seed dressing as liquid or powder, some formulations in combination with guazatine.

Guazatine
(di-(8-guanidino-octyl)amine)

Trade names 'MC 25', 'Panoctine', 'Guanoctine'
Properties White solid; slightly soluble in water.
Use Seed dressing usually at 0·8 g a.i. per kilogram of seed for cereals.
Caution Repellant to birds. Do not use treated seed for food.
Formulation Powder for seed dressing and aqueous solution.

Hexachlorobenzene

Properties A chlorinated hydrocarbon; colourless crystals; practically insoluble in water.
Use Selective fungicide used to control bunt of wheat by seed dressing.
Caution Some mammalian toxicity; can irritate skin. Do not use treated seed for food.
Formulation Usually as 10 per cent dust and with other seed protectants.

Hydroxyisoxazole (3-hydroxy-5-methylisoxazole)

Trade names 'Tachigaren', 'Hymexazole'
Properties Colourless crystals; moderately (8·5 per cent) soluble in water.
Use Soil and seed application to control many soil-borne pathogens. Also has growth-stimulating properties.
Caution Do not use treated seed for food.
Formulation Liquid concentrate and seed-dressing dust.

Quinacetol sulphate (di-(5-acetyl-8-hydroxyquinolinium) sulphate)

Trade names 'Fongoren', 'Risoter'
Properties Yellow crystalline; soluble in water. Low toxicity.
Use Protective fungicide against seed-borne pathogens. Used as a seed potato dip against *Rhizoctonia* spp.
Caution Do not use treated seed for food.
Formulation 80 per cent a.i. granules.

Quintozene (pentachloronitrobenzene (PCNB)

Trade names 'Brassicol', 'Tritisan', 'Terraclor'
Properties Colourless crystals; practically insoluble in water.
Uses Applied to soil and seed for control of Basidiomycete diseases, such as smuts, *Rhizoctonia* spp. and others.
Caution Do not use treated seed for food.
Formulation Usually as dusts.

Systemic fungicides

Since the late 1960s there has been a substantial development of systemic fungicides. These substances are absorbed by the plant and are translocated at least to some extent within the tissues. The efficiency of these substances does not depend on uniform coverage of the plant surface and they are often applied at low-volume or even ULV rates. They often have a therapeutic effect by eradicating disease lesions.

Most systemic fungicides move up the plant so that topical application to control root or stem-base diseases is not usually used. Application to seed and soil has achieved considerable success in the control of some diseases of foliage, stem and fruit. Many systemic fungicides have a narrower spectrum than the more widely used and conventional protectants. An additional problem is that strains of pathogens which are resistant to some systemic fungicides have been detected.

Aluminium *tris*(ethyl phosphonate)

Trade names 'Alliette', 'EPAL'
Properties White crystalline, solid soluble in water.
Use Systemic fungicide with particular activity against Oomycetes, can be translocated from shoots to roots.

Formulation 80 per cent wettable powder.

Benodanil
(2-iodobenzanilide (I))

Trade names 'Calirus', 'Mebenil' (a closely related compound)
Properties Crystalline solid.
Use Effective against Basidiomycetes such as rusts and *Rhizoctonia* spp. at about 0·2 per cent a.i.
Caution Some mammalian toxicity. Toxic to bees and fish.
Formulation Wettable powder and seed dressing for potatoes.

Benomyl
(methyl 1-(butylcarbamoyl)benzimidazol-2-ylcarbamate)

Trade name 'Benlate'
Properties White powder, with faint acrid odour; insoluble in water.
Use As a protectant and systemic fungicide active against a wide range of diseases including mildews and fruit rots, but excluding *Alternaria* spp. and *Cercospora* spp., rusts and Oomycetes. Used at about 0·05 per cent a.i. suspension for spraying. Pathogen tolerance has been reported after extensive use.
Caution Very low mammalian toxicity; safe to use after harvest.
Formulation Wettable powder (50 per cent).

Bupirimate
(5-butyl-2-ethylamino-6-methylpyrimidin-4-yl dimethylsulphamate)

Trade name 'Nimrod'
Properties Pale waxy solid; hardly soluble in water.
Use Systemic fungicide for the control of powdery mildews at 0·1 per cent a.i. on fruit and ornamentals.
Formulation An emulsifiable concentrate and dispersible powder.

Carbendazim
(methyl benzimidazol-2-ylcarbamate)

Trade names 'Derosal', 'Bavistin'
Properties Light grey powder; soluble in organic solvents but only slightly so in water.
Use Systemic fungicide controlling many fungal diseases of various crops. Pathogen tolerance has been reported after extensive use.
Caution Very low mammalian toxicity.
Formulation Wettable powder and liquid.

Carboxin
(2,3-dihydro-6-methyl-5-phenylcarbamoyl-1,4-oxathiin)

Trade name 'Vitavax'
Properties White powder, virtually insoluble in water.
Use Systemic fungicide active against Basidiomycete diseases such as smuts, *Rhizoctonia* spp., etc.
Caution Toxic to fish. Do not use treated seed for food.
Formulation Wettable powder and seed dressings. Some seed dressing formulations may contain poisonous organo-mercurial compounds (see p. 77).

Chloroneb
(1,4-dichloro-2,5-dimethoxybenzene)

Trade name 'Demosan'
Properties White powder with a musty odour; virtually insoluble in water.
Use Systemic fungicide for soil and seed application especially effective against *Rhizoctonia* spp. and *Pythium* spp., but less so to *Fusarium* spp. Rate depends upon seed, spacing, etc.
Caution Do not use treated seed for food.
Formulation Wettable powder.

Diclobutrazol
((2*R*, 3*R*)-and (2*S*, 3*S*)-1-(2,4-dichlorophenyl)-4,4-dimethyl-2-(1,2,4-triazol-1-yl)pentan-3-ol)

Trade name 'Vigil'
Properties Colourless crystalline solid, slightly soluble in water.
Use Broad-spectrum systemic fungicide with particular activity against rusts and powdery mildews.

Caution May be irritating to eyes.

Formulation 50 per cent and 12 per cent a.i. w/v suspension concentrate also formulated with carbendazim.

Dimethirimol
(5-butyl-2-dimethylamino-6-methylpyrimidin-4-ol)

Trade name 'Milcurb'

Properties White crystalline powder, slightly soluble in water.

Use Systemic fungicide for control of mildews on vegetables (especially cucurbits) and ornamentals applied to soil or foliage in dilute solutions. Pathogen tolerance has been reported after extensive use.

Caution Low mammalian toxicity.

Formulation Aqueous concentrate.

Dodemorph
(4-cyclododecyl-2,6-dimethylmorpholine)

Trade names 'BASF F238', 'Morpholine'

Properties Yellow liquid with characteristic odour; soluble in water.

Use Systemic fungicide used as an eradicant spray on fruit crops and some ornamentals. Used at about 10 per cent a.i. spray.

Caution

1 Low mammalian toxicity, but can irritate skin, eyes and nose.

2 Can be phytotoxic to some crops.

Formulation Liquid concentrate.

Ethirimol
(5-butyl-2-ethylamino-6-methylpyrimidin-4-ol)

Trade names 'Milstem', 'Milgo'

Properties White crystalline solid, slightly soluble in water.

Use Systemic fungicide used against cereal powdery mildews, absorbed by seedlings from seed dressing. Pathogen tolerance has been reported after extensive use.

Caution Treated seed not to be used for food.

Formulation Aqueous suspension for seed dressing and spraying.

Fuberidazole
(2-(2-furyl)benzimidazole (1))

Trade names 'Voronit', 'Furidazole'

Properties White powder, insoluble in water.

Use As seed dressing against *Fusarium* spp. diseases. Little effect against Oomycetes.

Caution Treated seed should not be used for food.

Formulation As a seed dressing, often in combination with other fungicides.

Furalaxyl
(methyl N-2, 6-dimethylphenyl-furoyl-(2)-alanate)

Trade name 'Fungarid'

Properties White crystalline solid, slightly soluble in water.

Use A protective fungicide with systemic properties, particularly active against Oomycetes.

Caution Slightly toxic to fish.

Formulation 50 per cent wettable powder.

IBP
(*S*-benzyl *00*-di-isopropil phosphorothioate)

Trade name 'Kitazin'

Properties A pale yellow powder, slightly soluble in water.

Use A systemic fungicide for the control of rice blast. Used at about 0·5 per cent a.i. Has insecticidal properties as well.

Formulation An emulsifiable concentrate, dust and granules.

Imazalil
(1-(allyloxy 2,4-dichlorophenethyl)imidazole)

Trade name 'Fungaflor'

Properties Light brown oily liquid, slightly soluble in water.

Use Broad-spectrum protectant fungicide against foliar diseases, and incorporated into seed dressings.

Formulation 20 per cent emulsifiable concentrate and incorporated with guazatine in liquid seed dressing 'Muridal'

Isoprothiolane
(di-isopropyl-1,3-dithiolan-2-ylidenemalonate)

Trade name 'Fuji-one'
Properties White crystals; slightly soluble in water.
Use A systemic fungicide active against rice blast.
Formulation Granules (12 per cent a.i.) Dust and emulsifiable concentrate.

Metalaxyl
((±)-methyl-2-[*N*-(2-methoxyacetyl)-2,6-xylidino] propionate)

Trade name 'Ridomil'
Properties Off white, crystalline solid, slightly soluble in water.
Use Systemic fungicide with particular activity against Oomycetes.
Caution May be irritating to skin and eyes.
Formulation 50 and 25 per cent a.i. wettable powders. 'Fubol' is a wettable powder mixture of 10 per cent metalaxyl and 48 per cent mancozeb.

Oxycarboxin
(2,3-dihydro-6-methyl-5-phenylcarbamoyl-1,4-oxathiin 4,4-dioxide)

Trade name 'Plantvax'
Properties White powder, slightly soluble in water.
Use Systemic fungicide active against Basidiomycete diseases such as rusts.
Caution Low mammalian toxicity. Toxic to fish.
Formulation Wettable powder and liquid concentrate.

Prothiocarb
(*S*-ethyl *N*-3-dimethylaminopropyl)thiocarbamate)

Trade name 'Previcur S70', 'Dynone'
Properties White crystalline; soluble in water.
Use A soil-applied systemic fungicide with specific action against Oomycetes. Used as a protectant.
Formulation As an aqueous solution of the hydrochloride, or soluble powder and granules.

Pyracarbolid
(3,4-dihydro-6-methylpyran-5-carboxanilide)

Trade name 'Sicarol'
Use Specific systemic activity against Basidiomycetes such as rusts, smuts and *Rhizoctonia* spp.
Formulation Wettable powder and dusts.

Pyrazophos
(*0*-6-ethoxycarbonyl-5-methylpyrazolo [1,5-*a*] pyrimidin-2-yl *00*-diethyl phosphorothioate)

Trade names 'Afugan', 'Curamil'
Properties Colourless crystals; slightly soluble in water.
Use Systemic fungicide active against powdery mildew on a wide range of crops at rates of 0·01 – 0·03 per cent a.i. foliar spray. Little root uptake.
Caution Moderate mammalian toxicity. Toxic to bees, fish and livestock. An anticholinesterase compound. Do not apply less than 2 weeks before harvest.
Formulation Emulsifiable concentrate and wettable powder.

Thiabendazole(TBZ)
(2-(thiazol-4-yl)benzimidazole

Trade names 'TECTO 40', '60' and '90' 'Mycozol'
Properties White colourless powder; slightly soluble in water.
Use Similar to benomyl — the active principle is similar.
Caution Very low mammalian toxicity, safe to use after harvest. Toxic to fish.
Formulation Wettable powders, emulsifiable concentrate and as fumigation tablets.

Thiophanate
(1,2-di-(3-ethoxycarbonyl-2-thioureido)benzene)
Thiophanate methyl (1,2-di-(3-methoxycarbonyl-2-thioureido)benzene)

Trade names 'Topsin', 'Cercobin', 'Mildothane', 'CYCOSIN'

Properties Crystalline solid; insoluble in water. Very low mammalian toxicity.
Use Protectant and systemic fungicide with similar range of action to the benzimidazole compounds, use as a 0·03 – 0·1 per cent a.i. suspension applied to foliage. Pathogen tolerance has been reported after extensive use.
Formulation Wettable powders and liquid concentrate.

Triadimefon
(1-(4-chlorophenoxy)-3,3-dimethyl- 1-(1,2,4-triazol-1-ylbutan-2-one))

Trade name 'Bayleton'
Properties Colourless solid, slightly soluble in water.
Use A systemic fungicide with protectant and curative activity against rusts and mildews of a whole range of crops. Used at 0·05 per cent a.i.
Caution Toxic to fish.
Formulation As wettable powders (5 and 25) per cent a.i.), emulsifiable concentrate and dust.

Tricyclazole
(5-methyl-[1,2,4]-triazolo [3,4-*b*]benzothiazole)

Trade name 'EL-291'
Properties Crystalline, soluble in water.
Use Systemic fungicide for rice blast disease control, applied to roots or foliage.
Caution Fairly low toxicity.
Formulation 75 per cent wettable powder, granules.

Tridemorph
(2,6-dimethyl-4-tridecyl-morpholine)

Trade name 'Calixin'
Properties Colourless liquid with slight odour, miscible with water.
Use Protectant and systemic fungicide used especially against cereal mildews and on fruit crops at low volume application of 1 – 2 per cent a.i. spray.
Caution Low mammalian toxicity; irritates eyes. Toxic to fish. Do not apply less than 2 weeks before harvest.
Formulation Liquid concentrate.

Triforine
(1,1′-piperazine-1,4-diyldi-[*N*-2,2,2-trichloroethyl)formamide])

Trade names 'Cela W524', 'Saprol'
Properties White odourless crystals, virtually insoluble in water.
Use Systemic fungicide effective against many foliar pathogens including mildews and rusts. Used at about 0·025 per cent a.i. as a foliar spray also suppresses red spider mites.
Caution Low mammalian toxicity.
Formulation Emulsifiable concentrate.

Antibiotics

Several antibiotics are used in crop protection, usually against specific diseases which warrant the use of these relatively expensive substances.

Blasticidin-S and **kasugamycin** have been promoted as antibiotics for specific control of rice blast (*Pyricularia oryzae*) and are available mainly in Japan.

Other anti-fungal antibiotics occasionally used for specialised purposes include griseofulvin and polyoxin.

Cycloheximide

Trade name 'Acti-dione'
Properties Colourless crystals; slightly soluble in water.
Use Applied highly diluted to certain high-value crops to control diseases such as mildews.
Caution Very toxic to animals.
Formulation As dilute solution and wettable powder.

Streptomycin

Trade names 'Agrimycin', 'Agristrep'

Properties White powder, soluble in water.
Use Antibacterial antibiotic used against fire blight on ornamentals and fruit and wildfire of tobacco.
Formulation As dilute dusts and wettable powders.

Tetracycline, Terramycin

Antibiotics such as oxytetracycline may be used occasionally in control of specific bacterial plant pathogens. They also have an effect on diseases caused by mycoplasmas and have been used experimentally in controlling these diseases.

Validamycin

Trade names 'Validacin'
Properties White powder, soluble in water.
Use Fungistatic to a narrow range of fungal diseases and used against rice sheath blight and other *Rhizoctonia* spp.
Caution Low mammalian toxicity.
Formulation Dilute solution or dust.

Nematicides

Most nematodes are soil-borne and can be controlled by chemicals applied to the soil. General sterilant chemicals used to control soil-borne diseases will also kill nematodes. Some of these have been mentioned earlier in this Chapter. In addition, other highly volatile organic halides which are less efficient as fungicides, can be injected into soil to obtain efficient nematode control.

Aldicarb (2-methyl-2-(methylthio)propionaldehyde *O*-methylcarbamoyloxime (I))

Trade name 'Temik'
Properties White odourless crystals, slightly soluble in water.
Use Applied to soil to control nematodes and some insect pests, it is systemic and absorbed by plants.
Caution An anticholinesterase compound (Part II substance). Protective clothing must be worn. Dangerous to fish, and wildlife.
Formulation As granules.

Carbofuran

See under insecticides (p. 131).

Dazomet

See under sterilants (p. 138).

Diamidafos (phenyl *N,N*-dimethylphosphorodiamidate)

Trade name 'Nellite'
Properties White odourless crystals, moderately soluble in water.
Use Specific nematicide to root-knot nematode larvae and used as a solution in irrigation water.
Formulation Water-soluble powder.

Dibromochloropropane (1,2-dibromo-3-chloropropane)

Trade names 'Fumazone', 'Nemagon'
Properties Brown liquid with mildly pungent odour, immiscible with water.
Use Applied to soil as drench or injection or by rotavating in granules. Soil needs to be warm ($> 20°C$) for best results. Perenial crops can tolerate fairly high doses.
Caution
1 Moderately toxic. Should be washed off skin immediately; irritant.
2 Phytotoxic to young crops.
3 Soil needs to be ventilated before crops are planted.
Formulation As emulsifiable concentrate and granules.

Diclofenthion (*O*-2,4-dichlorophenyl *O,O*-diethyl phosphorothioate)

Trade names 'VC13 Nemacide', 'Mobilawn'
Properties Colourless liquid, immiscible with water.

Use A non-systemic nematicide and insecticide, applied to soil at between 20 and 200 kg per hectare.

Caution Moderate mammalian toxicity; low phytotoxicity.

Formulation Emulsifiable concentrate and granules.

Dichloropropane-dichloropropene

Trade names 'D-D', 'Vidden D'

Properties A volatile amber liquid with a pungent odour.

Use Injected directly into soil at 200-400 kg per hectare and has a fumigant action. It has no harmful effect to soil flora and does not accumulate.

Caution

1 Moderately toxic. Irritant to skin and should be immediately washed off.
2 Phytotoxic.
3 Pre-planting interval at least 2 weeks.

Formulation As the undiluted chemical.

Ethoprophos
(*O*-ethyl *SS*-dipropyl phosphorodithioate)

Trade names 'Mocap', 'Prophos'

Properties Pale clear yellow liquid, slightly soluble in water.

Use A non-systemic, non-fumigant nematicide with acaricidal properties.

Caution A toxic chemical which should only be handled with appropriate protective clothing.

Formulation As granules and emulsifiable concentrates.

Fenamiphos
(ethyl-4-methylthio-m-tolylisopropylphosphoramidate (I))

Trade name 'Nemacur'

Properties Brown semi-solid; slightly soluble in water.

Use As a nematicide for soil application by broadcast, band application or root dips. Also has systemic activity against insects.

Caution Fairly toxic; use with caution and wear appropriate protective clothing.

Formulation As emulsifiable concentrate and granules.

Fensulfothion
(00-diethyl 0-4-methylsuphinylphenyl phosphorothioate)

Trade names 'Dasanit', 'Terracur P'

Properties Oily yellow liquid, slightly soluble in water.

Use Nematicide and insecticide with some systemic activity and long persistence, applied to foliage and soil.

Caution High toxicity; protective clothing should be worn.

Formulation Emulsifiable concentrate, dust and granules.

Isazophos
(*OO*-diethyl-*O*-[1-isopropyl-5-chloro-1,2,4-triazolyl-(-3)]phosphorothioate)

Trade name 'Miral'.

Properties A yellowish liquid slightly soluble in water.

Use A soil insecticide and nematicide.

Caution A fairly toxic chemical which should be handled with appropriate protective clothing.

Formulation As granules, emulsifiable concentrates and seed dressing.

Metham-sodium

See under sterilants (p. 139).

Methylisothiocyanate

See under sterilants (p. 139).

Oxamyl

See under insecticides (p. 132).

Tetrachlorothiophene

Trade name 'Penphene'

Properties A yellow-white solid; insoluble in water.

Use As a drench or injection into mineral soils for killing parasites and free-living nematodes.

Caution Pre-planting interval — 2 weeks. Moderately toxic to animals. Phytotoxic. Care needed when handling this chemical.

Formulation Emulsifiable concentrate.

Thionazin

See under insecticides (p. 128).

Further reading

Dekker, J. (1977). Chemotheraphy. In *Plant Diseases: An Advanced Treatise* Vol. 1. (eds Horsfall, J.G. & Cowling, E.B.) Academic Press: New York.

Dittmer, D.S. (ed.) (1959). *Handbook of toxicology* Vol. 5 *Fungicides*. Saunders: London.

Evans, E. (1968). *Plant diseases and their chemical control*. Blackwells: Oxford.

Goldberg, H.S. (ed.) (1959). *Antibiotics, their chemistry and non-medical uses. van* Nostrand: USA.

Hassell, K.A. (1969). *World Crop Protection*. Vol. 3. *Pesticides*. Iliffe Books: London.

Horsfall, J.G. (1956). *Principles of fungicidal action*. Chronica Botanica Co.: USA.

Lyr, H. (1977). Mechanisms of action of fungicides. In *Plant Diseases: An Advanced Treatise*, Vol. 1. (eds Horsfall, J.G. and Cowling, E.B.) Academic Press: New York.

Marsh, R.W. (ed.) (1977). *Systemic fungicides*. Longman: London.

Martin, H. (1964). *The scientific principles of crop protection*. 5th ed. E. Arnold: London.

Martin, H., Worthing, C.R. (eds) (1976). *Insecticide and fungicide handbook*. 5th ed. Blackwells: Oxford.

PANS Pesticide Index. (1979). PANS. Centre for Overseas Pest Research: London.

Rose, G.J. (1963). *Crop protection*. 2nd ed. Leonard Hill: London

Sharvelle, E.G. (1961). *The nature and uses of modern fungicides*. Burgess Publishing Co.: USA.

Siegel, M.R., Sisler, H.D. (eds) (1977). *Antifungal compounds*. Dekker: USA.

Thomson, W.T. (1978). *Agricultural chemicals IV. Fungicides*. Thompson Publications: USA

Torgeson, D.C. (ed.) (1967-8). *Fungicides, an advanced treatise*. Vol. 1. *Agricultural and industrial applications, Environment interactions*, Vol. 2. *Chemistry and physiology*. Academic Press: New York.

Wolfe, M.S. (1975). Pathogen response to fungicide use. *Proceedings of the 8th British Insecticide and Fungicide Conference 1975,* **3**, 813-822.

Worthing C. R. (ed.) (1979). *Pesticide Manual*. 6th ed. British Crop Protection Council: London.

Zwelg, G. (ed.) (1964). *Analytical methods for pesticides, plant growth regulators and food additives* Vol.3. *Fungicides, nematicides and soil fumigants*. Academic Press: New York.

Appendix
List of standard common names and alternative common and/or trade names for pesticides

Insecticides

Standard name	Alternative common and/or trade name
Biological compounds	
Bacillus thuringiensis	'Agritol', 'Bakthane', 'Biotrol', 'BTB–183', 'BTV', 'Dipel', 'Larvatrol', 'Thuricide', 'Tribactur', etc.
Carbamates	
Bufencarb	'Bux'
Carbaryl	'Carbaryl 85', 'Murvin', 'Pantrin', 'Septon', 'Sevin', et
Carbofuran	'CURATERR', 'Furadan', 'Yaltox'
Methiocarb	'Baysol', 'Draza', 'Mesurol', etc.
Methomyl	'Halvard', 'Lannate', 'Nudrin', etc.
Oxamyl	'Vydate'
Pirimicarb	'Aphox', 'Fernos', 'Pirimor', 'Rapid'
Promecarb	'Carbamult', 'Minacide'
Propoxur	'Baygon', 'Blattanex', 'Suncide', 'Unden', etc.
Chlorinated hydrocarbons	
Aldrin	'Aldrex', 'Aldrite', 'Drinox', 'Toxadrin', etc.
DDT	(many trade names and common names)
Dicofol	'Acarin', 'Kelthane', 'Mitigan', etc.
Dieldrin	'Alvit', 'Dieldrex,' 'Dilstan', 'Ensodil', 'Octalox', etc.
Endosulfan	'Cyclodan', 'Malix', 'Thifor', 'Thiodan', 'Thionex', etc.
Endrin	'Endrex', 'Hexadrin', 'Mendrin', 'Nendrin', etc.
Gamma-HCH	BCH, gamma-BCH, 'Lindane', etc.
Heptachlor	'Drinox', 'Heptamul', 'Velsicol', etc.
Mirex	'Dechlorane'
TDE	'DDD', 'DET', 'Rhothane', etc.
Tetradifon	'Duphar', 'Tedion V-18'
Tetrasul	'Animert', 'V-101'
Miscellaneous compounds	
Aluminium phosphide	'Celphos', 'Delicia Gastoxin', 'Detia', 'Phostoxin'
Bromomethane	'Dowfume MC', 'Embafume', 'Metafume', 'Profume', etc.
Copper(I) acetoarsenite	'Paris Green'
Dibromoethene	'Agrifume', 'Bromofume', 'Celmide', 'DB', 'Dowfume', 'EDB', 'Nemafume', 'Terrafume', etc.
Lead arsenate(V)	'Gypsine', 'Soprabel'
Mercury(I) chloride	'Calomel', 'Cylosan'

Natural organic compounds	
Bioallethrin	'Esbiol', 'D-Trans', 'Bioallethrine'
Nicotine	'Black Leaf 40', 'Nicofume'
Pyrethrins	
Rotenone	'Derris', 'Protex', 'Ro-Ko', 'Rotocide', 'Tubatoxin', etc.
Organic oils	
Petroleum oils	White oils
Tar oils	
Organophosphorus compounds	
Azinphos-methyl	'Benthion', 'Carfene', 'Gusathion' 'Super-Azin', 'X-Athion', etc.
Azinphos-methyl with demeton-S-methyl sulphone	'Gusathion MS', etc.
Carbophenothion	'Dagadip', 'Garrathion', 'Trithio', etc.
Chlorfenvinphos	'Birlane', 'Sapecron', 'Supona', 'Vinyphate', etc.
Demephion	'Cymetox', 'Pyracide', etc.
Demeton	'Solvirex', 'Systemox', 'Systox'
Demeton-S-methyl	'Demetox', 'Metasystox 55', etc.
Diazinon	'Basudin', 'DBD', 'Diazitol', 'Neocide', etc.
Dichlorvos	'Dedevap', 'Mafu', 'Nogos', 'Nuvan', 'Oko', 'Vapona', etc.
Dimefox	'Hanane', 'Terra-sytam', etc.
Dimethoate	'Cygon', 'Dantox', 'Rogor', 'Roxion', 'Trimeton', etc.
Disulfoton	'Disyston', 'Murvin 50', 'Parsolin', 'Solvirex', etc.
Ethion	'Embathion, 'Hylemox', 'Nialate', 'Rhodocide', etc.
Ethoate-methyl	'Fitios'
Fenitrothion	'Accothion', 'Agrothion', 'Dicofen', Folithion', 'Nuvanol', 'Sumithion', etc.
Fenthion	'Baycid', 'Baytex', 'Lebaycid', 'Queletex, 'Tiguvon', etc.
Fonofos	'Dyfonate', etc.
Formothion	'Alfix', 'Anthio', etc.
Malathion	'Malastan', 'Malathexo', 'Malathion', etc.
Mecarbam	'Afos', 'Murfotox', 'Pestan', etc.
Menazon	'Aphex', 'Saphicol', 'Saphizon', 'Sayfos', etc.
Methidathion	'Supracide', 'Ultracide', 'Phosfene'
Mevinphos	'Menite', 'Phosdrin', 'Phosfene'
Monocrotophos	'Azodrin', 'Monocron', 'Nuvacron', etc.
Naled	'Bromex', 'Dibrom', etc.
Omethoate	'Folimate'
Oxydemeton-methyl	'Metasystox-R', etc.
Oxydisulfoton	'Disyston-S', etc.
Parathion	'Bladan', 'Fosferno', 'Niran', 'Thiophos'
Parathion-methyl	'Azofos', 'Bladan M', 'Dalf', 'Folidol-M', 'Metacide', 'Metafos', 'Nitrox-80' 'Wofatox', etc.
Phenisobromolate	'Acarol', 'Necron', etc.
Phenthoate	'Cidial', 'Elsan', 'Papthion', 'Tanone', etc.

Phorate	'Granutox', 'Rampart', 'Thimet', 'Timet'
Phosalone	'Embacide', 'Rubitox', 'Zolone'
Phosmet	'Appa', 'Germisan', 'Ginicide', 'Imidan', 'Prolate', etc.
Phosphamidon	'Dicron', 'Dimecron', 'Dovip', 'Famfos'
Phoxim	'Baythion', 'Valexon', 'Volaton', etc.
Pirimiphos-ethyl	'Fernex', 'Primicid', 'Primotec'
Pirimiphos-methyl	'Actellic', 'Actellifog', 'Blex'
Prothoate	'Fac', 'Fostin', 'Oleofac', 'Telefos'
Quinomethionate	'Erade', 'Forstan', 'Morestan', etc.
Schradan	'Pestox 3', 'Sytam', etc.
TEPP	'Bladen', 'Fosvex', 'Nifos T', 'Tetron', 'Vapotone', etc.
Tetrachlorvinphos	'Gardona', 'Ostabil', 'Rabon', 'Ravap'
Thiometon	'Ekatin', 'Intrathion', etc.
Thionazin	'Nemafos', 'Nemasol', 'Zinophos', etc.
Thioquinox	'Eradex', 'Eraditon'
Trichloronate	'Agritox'. 'Agrisil', 'Fenophosphon', 'Phytosol', etc.
Trichlorphon	'Anthon', 'Chlorofos', 'Danex', 'Dipterex', 'Notox', 'Tugon', etc.
Tricyclohexyltin hydroxide (Cyhexatin)	'Plictran'
Vamidothion	'Kilval', 'Rhodiamide', 'Trucidol', 'Vamidoate', 'Vation'

Substituted phenols	
Binapacryl	'Acricid', 'Ambox', 'Dapacryl', 'Endosan', 'Morocide', etc.
DNOC	'Cresofin', 'Detal', 'Dinitrol', 'Sandolin', 'Sinox', 'Triforcide', etc.
Pentachlorophenol	'Dowicide', 'Pentacide', 'Permacide', 'Santobrite', 'Santophen', 'Timbertox'

Chemicals used to control plant diseases

Inorganic	
Bordeaux mixture	
Burgundy mixture	'Burcop', etc.
Cheshunt compound	
Copper(II) hydroxide	'Kocide'
Copper naphthenates	'Cuprinol'
Copper oxychloride	'Blitox', 'Cupravit-Forte', 'Vitigran', etc.
Copper(II) sulphate(VI)	'Blue stone', 'Blue vitriol'
Copper(I) oxide	'Copper-Sandoz', 'Cuprocide', 'Perenox'
Lime-sulphur	
Oxine copper	Copper hydroxyquinolate

Sulphur (brimstone)	
Mercury(I) chloride	calomel
Mercury(II) chloride	corrosive sublimate
Methoxymethyl mercury salts	'Aretan', 'Ceresan'
Phenylmercury acetate and chloride (PMA, PMC)	'Agrosan D', 'Ceresol', 'Leytosan'
Other organo-mercurial compounds	'Ceresan M', 'Granosan M', 'Panogen', 'Tillex'
Triphenyltin compounds	'Brestan', 'Du-Ter', 'Stannoram'
Methylarsinic sulphide	'MAS', 'Rhizoctol', 'Urbasulf'
Organic compounds	
General protectants — dothiocarbamates	
Cufraneb	'Macuprax'
Ferbam	'Fermate'
Mancozeb	'Dikar', 'Dithane M-45'
Maneb	'Dithane M-22', 'Manzate'
Metiram	'Polyram-Combi'
Propineb	'Antracol'
Thiram	'Arasan', 'Nomersan', 'Tersan'
Zineb	'Dithane Z-78', 'Polyram-Z'
Ziram	'Cuman', 'Milbam', 'Zerlate'
General protectants — other compounds	
Anilazine	'Dyrene', 'Triazine'
Captafol	'Ortho-Difolatan'
Captan	'Orthocide'
Chlorthalonil	'Bravo', 'Daconil 2787', 'Termil'
Dichlofluanid	'Elvaron', 'Euparen'
Ditalimphos	'Plondrel'
Dithianon	'Delan'
Dodine	'Cyprex', 'Melprex'
Drazoxolon	'Ganocide', 'Mil-Col'
Fenarimol	'El-222', 'Rimidin'
Folpet	'Phaltan'
Halacrinate and captafol	'Tilt'
Pyridinitril	'Ciluan'
Tolylfluanid	'Euparen M'
Urbacid	'Tuzet'
Protectant fungicides for seed and soil application	
Bronopol	'Bronocot'
Chloranil	'Spergon'
Dichlone	'Phygon', 'Uniroyal USR 604'
Etridiazole	'Terrazole', 'Ethazol'
Fenaminosulf	'Dexon'
Fenfuram	'Panoram'
Guazatine	'MC 25', 'Guanoctine'

Hexachlorobenzene	
Hydroxyisoxazole	'Tachigaren'
Quinacetol sulphate	'Fongoren', 'Risoter'
Quintozene	'Brassicol', 'Terraclor', 'Tri-PCNB', 'Tritisan'

Specific narrow-range organic protectant fungicides

Binapacryl	'Acricid', 'Endosan', 'Morocide'
Chlorquinox	'Lucel'
Dicloran	'Allisan', 'Botran'
Dinobuton	'Talan', 'Acrex'
Dinocap	'Karathane', 'Crotothane'
Edifenphos	'Hinosan'
Fluotrimazole	'Persulon'
Iprodione	'ROVRAL'
Nitrothal-isopropyl	'Kumulan', 'Pallinal'
Petroleum oils	
Phenylphenol	'Dowicide 1', 'Dowicide 2'
Procymidone	'Folosan', 'Fusarex'
Quinomethionate	'Morestan'
Tecnazene	'Sumisclex'
Thioquinox	'Eradex'
Vinclozolin	'Ronilan'

Systemic fungicides

Benodanil	'Calirus'
Benomyl	'Benlate'
Bupirimate	'Nimrod'
Carbendazim	'Bavistin', 'Derosal'
Carboxin	'Vitavax'
Chloroneb	'Demosan'
Dimethirimol	'Milcurb'
Dodemorph	'Morpholine' 'Meltatox'
Ethirimol	'Milstem'
Fuberidazole	'Voronit'
Furalaxyl	'Fongarid'
IBP	'Kitazin P'
Imazalil	'Fungaflor'
Isoprothiolane	'Fuji-one'
Metalaxul	'Ridomil'
Oxycarboxin	'Plantvax'
Prothiocarb	'Dynone', 'Previcur'
Pyracarbolid	'Sicarol'
Pyrazophos	'Afugan', 'Curamil'
Thiabendazole	'Mycozol', TBZ, 'Tecto'
Thiophanate and thiophanate-methyl	'Cercobin', 'Mildothane', 'Topsin'
Triadimefon	'Bayleton'

Tricyclazole	'El-291'
Tridemorph	'Calixin'
Triforine	'Cela W 524'

Antibiotics	
Blasticidin-S	'Bla-S'
Cycloheximide	'Acti-dione'
Kasugamycin	'Kasumin'
Streptomycin	'Agrimycin 100', 'Agri-strep'
Tetracycline	
Validamycin	'Validacin'

Nematicides	
Aldicarb	'Temik'
Carbofuran	'Furadan'
Dazomet	'Bazamid', 'Mylone'
Diamidafos	'Nellite'
Dibromochloropropane	'Fumazone', 'Nemagon'
Diclofenthion	'Nemacide VC-13'
Dichloropropane-dichloropropene	'D-D', 'Vidden D'
Ethoprophos	'Mocap'
Fenamiphos	'Nemacur'
Fensulfothion	'Dasanit', 'Terracur P'
Isazophos	'Miral'
Metham-sodium	'Vapam'
Methyl isothiocyanate	'Trapex'
Methyl isothiocyanate and dichloropropane-dichloropropene	'Vorlex'
Oxamyl	'Vydate'
Tetrachlorothiophene	'Penphene'
Thionazin	'Nemafos', 'Zinophos'

Sterilants	
Chloropicrin	
Dazomet	'Basamid', 'Mylone'
Dibromoethene	'Bromofume', 'Dowfume W-85'
Dibromoethene and dichloropropene	'Darlone'
Methanal (formaldehyde)	'Formalin'
Metham-sodium	'Vapam', 'VPM'
Methyl bromide	'Dowfume MC-2'
Methyl isothiocyanate	'Trapex'
Methyl isothiocyanate and dichloropropane-dichloropropene	'Vorlex'

Index